ADVAN(
INSIDE COMMERCI/

"Watch what happens when the guy sitting next to you reads this book—and you don't."

Mouji Linarez, Florida Healthcare Market Leader
DPR Construction

"Read it—and don't look back."

Brandice Masse Schaefer, Senior Manager
Gilbane

"I encourage everyone I place to read *Inside Commercial Construction's MVPs*—to find out why every company I recruit for would hire Coty Fournier in a second."

Mike Kittelson, President
Trinity Search Group, Commercial Construction

"The only thing better than Coty's book is her live on stage—go see her! Until then, this book is the next best thing to a one-on-one coaching conversation with her."

Richard Johnson, President
The Blue Book Building & Construction Network

"*Inside Commercial Construction's MVPs* is jamb-packed full of powerful insights. It's a provocative assessment of the industry and a foundation to differentiate yourself from the masses."

Jonathan Antevy, Co-founder
e-Builder

"On one hand, it's a must-read for construction management rookies looking for the fastest path to the top. On the other hand, it's one hell of a wake-up call for many seasoned veterans."

Harley Miller, President
Miller Construction Company

"Coty Fournier should be the official ambassador for the U.S. Commercial Construction Industry—to tell it like it is and raise the bar even higher for construction leaders everywhere."

Beverly Raphael, President & CEO
RCC Associates, General Contractor

"Talk about blazing a new path and leaving a trail for others— Coty Fournier has written the absolute best book on construction career advice I've ever seen. No one talks like this. No one writes like this. Finally, someone got it right."

Neil Hammack, Vice President of Operations
Kaufman Lynn Construction

"You're holding the keys to your future in the commercial construction industry, right here in your hands."

Terry Phillips, Executive Director
Allied Construction Industries

"Honest. Provocative. Smacks you in the face with real world career guidance. Ignore this book at your own risk."

Craig Noble, Director of Marketing
Thermal Concepts
and Co-founder of Jobsite123.com

INSIDE COMMERCIAL CONSTRUCTION'S MVPs

INSIDE
COMMERCIAL
CONSTRUCTION'S

MVPs

7
reasons
why they get
promoted faster,
make more money, and
enjoy a seemingly unfair
advantage over everybody else.

COTY LEIGH FOURNIER

HUNTING HOUSE
Fort Lauderdale, Florida

Published by Hunting House
Fort Lauderdale, Florida

Cover Image: © Baitong333 www.fotosearch.com Stock Photography

ISBN: 978-0-9908584-0-9
Library of Congress Control Number: 2014917292

First Hunting House printing: October 2014

TABLE OF CONTENTS

AUTHOR'S NOTE

Attention readers: Please excuse the primary use of nouns and pronouns in the masculine form. You will see the words "guy" and "guys" and "he" and "him" used predominantly throughout the book. This is done purely for the sake of ease in reading, to avoid the cumbersome nature of "he/she" or "him/her" and other similar conventions. There is absolutely no intent to offend or exclude women readers, and the author proudly acknowledges the growing number of women making incredible contributions to the commercial construction industry every day.

IN APPRECIATION

I'm profoundly grateful for all of the people who've helped me in my career. Although the words in this book are my own, much of their advice and wisdom are laced together with mine. The commercial construction industry could've quickly chewed me up and spit me out—but it didn't. I survived and accomplished more than most would have predicted. Some of that credit I will keep, but much of it goes to talented and special people that I was lucky to be around and privileged to be mentored by along the way. There are too many to name, but I carry all of them with me as I embark upon every new challenge. If you're wondering whether or not your name is on my gratitude list—it is.

An extra big thank you is due to everyone who took chances on me when I was green, and focused on my potential when that was pretty much all I had to offer. Thank you to those who lifted me up, dusted me off, and forgave me when I made mistakes or just plain didn't know what the hell I was doing. Thank you to everyone who was patient enough to answer my thousands of questions, because that's how I like to learn. And thank you to those who invested their own time, energy and money in me and my ideas, both good and bad.

Lastly, a long distance shout-out to the vast array of wicked smart and inspirational people that I've never met, but I admire from afar, inside our industry and out. I love to learn from other people's stories and accomplishments. Much of what I've learned, I learned from watching others who didn't even know I was watching. I highly recommend it.

In closing, I do declare that I'm filled with joy and appreciation for my family and all of my fantastic friends and loved ones who encouraged me to write this book. Thank you for your patience, support and the hours and hours of listening, reading, editing and feedback. I love you all.

I've done a few good things in the world, but nothing compares to how proud I am to be the daughter of Paul and Joyce Fournier, the sister of Nicole and Brandon Fournier, an aunt to Norah McAteer Fournier, and the mother of Brielle Nicole Fernandez, my one and only amazing daughter. I hope that all of the professional challenges I have undertaken in my life will encourage her to follow her own dreams.

She once called me a Feisty Nerd because I've been studying theoretical physics, particle physics and cosmology with extreme passion for my entire adult life. I'm madly in love with all things fundamental to the universe, and I could just burst with fascination that we all belong to a species of life capable of contemplating itself and daring enough to solve the grandest of all mysteries—on the largest and smallest of scales. I just can't get enough of it.

I've been working diligently ever since to live up to that awesome nickname. I think it's a fantastic thing to be. Thank you, Brielle, for showing me who I really am in all the ways that count.

Coty

INTRODUCTION

This is a book about the MVPs in the commercial construction industry. Not the most valuable companies, the most valuable people. It's not about names, because names are distracting. It's about how the MVPs think, what they know, what they focus on, and, most importantly, what they do. You'll also find a little bit of what they talk about behind closed doors and some of what doesn't get talked about at all.

I wrote this book because it's the book I wish I'd had throughout my own career—with all the stuff they cannot teach you in school and the golden nuggets that many top-performers are either unable to articulate or unwilling to share. Access to a book like that would've opened my eyes to important concepts sooner and made some of my difficult career choices easier. So after years of respected colleagues encouraging me to step up and write *that* book, I decided to take on the challenge and pave an easier way for others.

In this book, you'll find some of the theories and strategies I've developed over the past twenty-five years of working for general contractors, project owners, and construction technology companies—and many insights gleaned from others with exceptionally fast-track careers and impressive compensation packages. You could say that I've been respectfully spying for a long, long time on highly successful construction executives who are looking back at their colleagues and competitors in the rear view mirror. In short, this book contains seven ways to go further and faster and make a lot more money than the guy sitting next to you . . . or the guy down the street.

As I speak around the country on the strategies included in this book, I'm often asked to describe each chapter in a nutshell, to sum up the most important messages in a sentence or two. That's never easy, but here goes.

CHAPTER 1:
WHAT ELSE YOU GOT?

The opening chapter will shock a few people. It's a serious heads up on the commoditization of the industry—including all of us who work in it—and how the MVPs are winning the differentiation game.

CHAPTER 2:
THE TRAILER AIN'T THE FIELD

Chapter two is all about defining actual field experience—what it is and what it isn't. It also gives you a sneak peek into some of what the MVPs take away from their time in the field, and the real reason why paying your dues out there is drop-dead essential to your career.

CHAPTER 3:
GET YOUR OWN JERRY MAGUIRE

This strategy takes an entertaining and enlightening look at the similarities between sports agents and executive search professionals, and explains how the MVPs create leverage in the marketplace with a sometimes controversial, but highly effective, silver bullet.

CHAPTER 4:
BEWARE OF THE PROJECT MANAGER MYTH

The project manager myth is intended to be an eye-opener for commercial construction professionals charging down the project management path. It explores the price you may end up paying for it, and why so many of the MVPs choose to leap frog right over it.

CHAPTER 5:
BE THE "I" IN TEAM

This one may seem like a politically incorrect philosophy on teamwork, but it's really about how the MVPs earn the respect and admiration of their teammates and command their rightful place in the spotlight.

CHAPTER 6:
PEOPLE ARE NOT SHOP DRAWINGS

Chapter six offers a compelling theory to explain why we all have a little voice inside our heads—I call mine Pistol Pete—that get us into trouble. You'll learn how and why the MVPs are ignoring him all the way to the top.

CHAPTER 7:
RIGHT, RIGHT, RIGHT, LEFT, RIGHT

The final chapter illuminates why lasting success in construction management no longer stems from excellence in construction management skills and reveals the most sought-after ingredient in the MVPs' secret sauce.

Some say my ideas are provocative and a little hyper-progressive—and I certainly hope so. I despise small talk and baby steps. Each one of these chapters is intended to make you think about our industry in a radically different way; however, I don't expect everyone to agree with me on everything. That's neither likely nor the real point.

What I do hope is that you'll put your own stamp on what you've read and do something about it. This book is really about encouraging you to take a long hard look at everything that's happening around you, and question all of it—even the good stuff—so you can formulate your own theories about how the industry actually works, and then boldly reimagine how it *could* work. From there you can develop informed strategies to make your own personal dent in the construction universe.

Every successful change agent will tell you that questioning and dissatisfaction with the status quo is not disrespectful; it's the first step toward reinvention and revolution. Nothing improves when everything is deemed "good enough" to carry on with business as usual. Regardless of your age, experience level, title, or position on your company's organizational chart, you can use this book to start building the iconic career and life of your dreams—and set an inspiring example for others to follow.

When you finish reading, if you're looking for some instant gratification or a place to focus your energies first, try going back to chapter six. Of all the strategies, it may contain the lowest-hanging fruit. The behavioral insights are intended to create immediate and high collective gains for all construction professionals, the companies they work for, and the industry as a whole. By shining a light on this epidemic behavior, and giving it a name, we can all stop engaging our Pistol Petes inappropriately, encourage others to do the same, and start a movement.

Truth be told, this book was almost never written or published because of my own Pistol Pete. He can shoot holes in just about everything you're about to read—and he has. I've been dueling with him for years, completely out-gunned most of the time. That's why we teach what we most need to learn, and those who teach learn the most.

Enjoy it.

Profit from it.

Pay it forward.

Coty

HEADS UP:

Do not implement any of the strategies in this book
unless you are prepared to advance your career,
make more money, and ruffle a few feathers.

All three will happen.

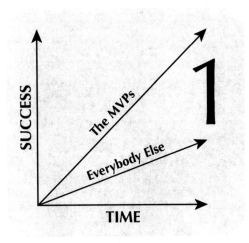

WHAT ELSE YOU GOT?

Commercial construction professionals who are promoted from operations to business development learn one of the most sobering, yet also empowering, truths about the industry. Unfortunately, very few people get that promotion, so they never learn this truth, or they learn it too late. It goes something like this:

> You can take virtually any contractor's sales materials and presentation content and swap it out with the same from their closest direct competitors and no one will know the difference.

What!
Really?
Yes.

No matter how much time and agonizing effort most companies spend on these monumentally important activities, in the
end the resulting pitches all sound pretty much identical to the
listening ear. Everyone claims to have the same five qualifications
floating amidst a sea of well-intended polish and decoration. But
when you boil it all down, here's what left:

> We deliver (1) quality construction projects, (2) on
> time and (3) on budget, and here are some (4) client
> testimonials from projects similar to yours to prove it.
> How do we do it? We have the (5) best people.

Some companies use few words that are short and to the point.
Some use lots of words, case studies, cool graphics, and other
eye-catchers. But for all intents and purposes, the only differences
in the sales and marketing messages between one contractor and
another are their logos, pictures, color choices, and who can do
what with PowerPoint. Each company is essentially saying, *"Pick
us! Pick us! We're the best!"* while defending that claim in much
the same manner as the next company. And so on and so on.

Nobody likes to hear this. It can be deflating and frustrating.
Especially if you've worked in the business development or marketing arena, and therefore know how hard it is to differentiate
your company's qualifications from your competitors in any believable fashion. But if you accept this truth and work with it,
instead of arguing and wishing it weren't true, you can capitalize
on it. The MVPs most certainly do. They use it to help the companies they work for—and fast-track their own careers at the same
time.

Exploring the root of this truth in further detail is enlightening. You'll soon see why it applies directly to you and your career, and how you can use it to your advantage.

DON'T BET THE MINIMUM ANTE

For lack of a better word, let's call those five qualifications the *basic* ones because they've become a common denominator among all legitimate contractors. Those very things that all contractors give their lives to achieve are now largely perceived to be the minimum ante required to stay in the game. For the most part, the basic qualifications are now expected to even make the shortlist for consideration and therefore no longer play much of a role in the decision-making process.

This expectation leaves project owners sifting through nearly every contractor's Request for Qualifications (RFQ) binder wondering, *"Yeah, yeah, yeah. What else you got?"* and sitting through their sales presentations feeling like they're watching a movie they've seen several times before. When all companies are pushing the same basic qualifications, it gets repetitive. More to the point, it isn't helpful. The minimum ante works fairly well when comparing apples to oranges, but it fails when comparing apples to apples. Decision-makers want and need one of those apples to stand out in order to help guide and justify their decision. They're quite literally in desperate search of tangible differentiation among all of the presenting companies—something that rings logically or emotionally true and valuable to the project at hand. That's hard to find these days in the minimum ante. Project owners are often left wondering which contractor can bring something *else* to the table, a company who can further reduce the known risks, save time or money for reinvestment, or somehow increase the likelihood of the project's success.

A handful of companies are rising to the challenge, becoming increasingly successful at the differentiation game—but most companies are not. They don't do a good job of making their differentiators believable or articulating the actual value proposition of the differentiators they're claiming to have. Worse yet, some don't have any real differentiators at all, and that is the saddest root of the truth.

LOOK IN THE MIRROR

Now twist the point and get personal. All of this applies to you and your career because everything translates directly from companies to individuals. Commercial construction professionals are in competition with one another for positions, promotions, and opportunities in much the same way that construction companies are in competition with one another for clients and projects. The landscape is so competitive that a reflective truth has emerged:

> You can take virtually any construction professional's resume or listen to the way he describes his talents, strengths, and relevant experience in a job interview and swap that out with the same from one of his legitimate competitors and no one will know the difference.

Recruiters, hiring managers, and senior executives often struggle to discern meaningful differences in the potential short- and long-term value of Construction Professional X, Y, or Z—when those people are roughly apples-to-apples candidates. They may see a small difference in the number of projects each candidate has worked on, or the relative size of those projects, but they won't likely perceive there to be a significant difference in their basic qualifications. Which are? Of course you already know. Virtually all legitimately qualified construction professionals applying for a certain position on a project team will claim to have the same five basic qualifications as the next guy, and his resume and interview pitch will sound something like this:

> *I have a strong track record for building (1) quality construction projects, (2) on time and (3) on budget, and here are my (4) references to prove it. I am the (5) best person for the job.*

This duplicity is certainly not limited to project managers, engineers, and superintendents. You can extrapolate the point and the same will hold true for estimators, preconstruction managers, business development managers, senior executives, and everyone else on the payroll in any sizable commercial construction company. All positions have a minimum ante required to play—and a certain set of basic qualifications that are required to do the job well and be considered successful. Whatever those basic qualifications are, you can safely assume that your legitimate competitors have them and some will have them in spades.

At this point, you've probably already deduced the second key assumption. Everyone involved in making a decision about whether or not to hire or promote you into a specific position already expects you to have that position's basic qualifications in the bag; therefore, what they're really doing is grasping for a definitive answer to their most pressing question: *"What else you got?"*

The market forces that are driving Construction Companies A, B, and C into the meaningful differentiation game are also driving Construction Professionals X, Y, and Z into the same game. The only difference between the company game and the personal game is awareness. Most people are aware that the company level game exists and therefore recognize that you have to play it wisely to win clients and projects. They may not yet play it well or consistently, but they'll concede that playing the game is required if you want any chance of ever getting work on something other than lowest qualified bid. Conversely, few people realize that you have to play the same game to advance your own career. So they never step in front of the mirror and honestly ask themselves: *"What else do I got?"*

Meanwhile, guess who does? The MVPs are standing in front of the mirror all the time, completely aware of the game that's being played and what's required to win. While the average person continues jabbing right-handed with the basic qualifications, the MVPs switch to southpaw mid-way through the third or fourth round and throw in what else they got. Half the time, their competitors don't even know what hit 'em.

BET YOUR BONUS QUALIFICATIONS

The MVPs in the commercial construction industry have a rock solid answer to the *"What else you got?"* question teed up and ready to go on every aspect of their entire career. In fact, they have several answers, strengthened by the ability to articulate each unique value proposition through powerful storytelling. Their answers demonstrate valuable skills and talents and relationships that stretch far beyond the basic qualifications expected for any particular position, into the *bonus* qualifications that seal the deal. They know how to sell themselves in all the ways that count and there's nothing fake or underhanded about it. Companies buy what they're selling because their bonus qualifications are real, valuable, and beautifully articulated.

Bonus qualifications can be any number of things in the commercial construction industry. Like a special talent—particularly if it's a rare talent—or positive relationships with clients and industry influencers. They could also be something a bit more intangible, like a unique attitude or a demonstrated willingness to handle things that others find difficult. You have to think outside the proverbial construction box, beyond the basic qualifications that everyone is traditionally focused on, and (1) identify your unique strengths and capabilities and (2) start showcasing the ones that can be turned into a valuable asset for your company.

Chew on these three delicious and bankable examples. They illustrate the wide spectrum of bonus qualifications in the marketplace and why the commercial construction industry needs them. Be inspired to think broadly and aim high as you identify and develop your own.

Example 1:
Q: What else you got?
A: Relationships that bring in work.

The MVPs work very hard to nurture positive relationships with all project owners and their representatives throughout their careers, and those relationships are leveraged to benefit the companies they work for. Some have strong enough client and industry relationships to establish client pull—meaning they have the potential to bring in more work. Here are four common scenarios and the resulting benefits to everyone involved.

1. Meet Frank Facilities. He's a construction manager at Healthy Hospital. Frank has worked with Good Guys Construction on his last two renovation projects and he's happy with the results. Frank credits the success of the projects to his positive working relationship with Good Guys Construction, and, in particular, his relationship with Peter Project Manager, who served as the project manager for Good Guys Construction on both projects. Frank has a very high degree of trust in Peter, because Peter made Frank's job easier and less stressful on project No. 1 and that played a pivotal role in Good Guys Construction being awarded project No. 2. As you might guess, Frank specifically requested that Peter serve as his project manager again.

    RESULTING BENEFIT: Peter's relationship with Frank at Healthy Hospital is a bonus qualification for Peter and a highly valuable asset for Good Guys Construction. This is especially true if Healthy Hospital has a robust 5-year development plan and more construction projects on the horizon.

2. Let's continue on with Frank Facilities and Peter Project Manager a little further. Peter was recently lured away from Good Guys Construction by another general contractor that does healthcare work called Done Right Construction. When Frank finds out that his favorite project manager now works for Done Right Construction, the door opens for Done Right Construction to begin working for Healthy Hospital because Frank's loyalty to Peter is greater than his loyalty to Good Guys Construction. Frank basically wants to work with Peter no matter what company Peter is working for because—all things being relatively equal, and they often are—Frank would prefer to stick with the guy who has already proven to make his job easier.

RESULTING BENEFIT: Peter's relationship with Frank at Healthy Hospital is becoming even more valuable and is now an asset for Done Right Construction. You get the idea.

3. One more scenario with Frank Facilities and Peter Project Manager also applies. If Frank leaves his job at Healthy Hospital and goes to work for Doctor Depot, he will likely bring opportunities for Peter Project Manager and Done Right Construction right along with him.

RESULTING BENEFIT: Peter's relationship with Frank has become even more valuable because his company now has a good chance of picking up work from Doctor Depot.

4. Lastly, let's change our cast of characters. Meet Steve Sales. Steve is vice president of business development for Beautiful Builders. Steve has nurtured a close friendship over the years with Mike Maintenance, who sits on the selection committee for design and construction activities at Super Schools. Steve and Mike go way back and Mike trusts Steve's opinion implicitly (even though Mike normally hates sales people) because Steve never lies to him. Steve always tells Mike the truth about whether or not Beautiful Builders is the best contractor choice for any given project at Super Schools. In fact, Mike's favorite story to tell about Steve is that once when Mike asked Steve why Beautiful Builders wasn't going after the new high school expansion project coming out, Steve told him:

> We're not going after it because the only two guys we have available to run it right now are unqualified to do it. Actually, one of them is qualified, but he's really tough to get along with and I don't think you would like him. I'm afraid he's not the right fit. Look, Mike, when we have the right people available for a project, we go for it. When we don't, we don't—because I need you to believe me when we do.

RESULTING BENEFIT: Beautiful Builders gets Mike's vote virtually every time they compete for a project at Super Schools because his trust level is high. Mike also has powers of persuasion over his colleagues on the selection committee, which results in Beautiful Builders winning a few projects at Super Schools every year. Steve's relationship with Mike is clearly a bonus qualification and it is likely to remain intact, no matter what company Steve works for or what school system Mike works for—and when either of them changes employers the relationship follows and potentially grows in value.

Scenarios like these happen all the time and they're impressive. In short, Peter Project Manager and Steve Sales are money in the bank for the companies that employ them. If you have relationships like these, make sure your boss and everyone above him knows it. If you don't yet have clients who'll go to bat for you like the examples outlined above, this is your heads-up to start nurturing them so you can compete with the MVPs who do.

Example 2:
Q: What else you got?
A: Teaching skills to train other people.

One of the most popular topics people love to argue and complain about in the commercial construction industry is training. Nearly everyone has thrown in his or her two cents at one time or another, while hypocritically running up and down both sides of the fence. They go from needing and wanting training to dreading and forgetting it. Half the people who complain about lack of training on X will turn around and complain when asked to spend time training someone else on Y. Or they'll beg for training, get it, and complain all the way through it. Worse yet, they may be asked to sit through training they really do need, but don't really want, so they show up mentally absent—and all kinds of other dysfunctional nonsense.

Some people are so emotionally fed up with the entire training dilemma that they've become apathetic. They like the idea of it on the surface, but they rarely see it work, so they're unconvinced that it's ever time well spent. The general apathy is understandable and everyone's frustrations are warranted because, in order for training outcomes to improve, there are real challenges that need to be overcome, such as these:

1. Almost everything involved in the construction of a commercial building is incredibly hard to teach—and ironically—learning all of the technical stuff is child's play compared to learning how to navigate the psycho-dynamics of a jobsite and everyone's conflicting agendas. It's an understatement to say that you cannot learn everything you need to know about the commercial construction industry from a book or classroom setting.

2. It takes a long, long time to teach and train people properly. Delegation, mentoring, and the transfer of knowledge from veteran construction professionals to rookies are luxuries our industry begs to afford. The short-term expense is often hard to justify against the long-term gain because time is very expensive in construction.

3. Great teachers and trainers are very hard to find. They're rare in any industry on virtually any subject. Teaching requires knowledge and relevant experience, plus a variety of other talents that most people don't have, including the ability to inform, inspire, entertain, and engage. At its essence, teaching is a performance art.

All of these things make it difficult to effectively train people in the construction industry. We have very few good teachers at our disposal. The stuff we need to teach is very hard to teach, and we have virtually no time to teach it. None of this is good news. So the fastest and easiest solution for most companies is to find a few needles in the haystack—meaning highly knowledgeable and insightful construction professionals who can actually teach—and then exploit their skills.

Imagine being great at what you do, coupled with the ability to teach it to others. Owners and senior executives pine for this. When they sit around the boardroom table celebrating the success of one of their star players, someone always says, *"Man, I wish we could clone that guy."* Teaching others what you know, and how to do what you do, are the closest things to cloning yourself, and the more proactively you do it, the better.

Waiting to be asked to teach something is nowhere near as impressive as spotting the need and offering to solve the problem without being asked. When the MVPs see their colleagues needing help with something, they share what they know, and they remain forever mindful to train their potential replacements in order to free themselves up for promotion when the timing is right.

If you can teach, step up to the plate, and claim your rightful throne at the head of the class, whether it's a class of many or privately one-on-one.

Example 3:
Q: What else you got?
A: Command of the stage.

Virtually all companies in the commercial construction industry have to make formal sales presentations in the business development process. It cannot be avoided if you want to grow your client base or simply maintain the one you already have. Participation in sales presentations causes tremendous anxiety for the vast majority of all people, in any industry, and that anxiety often causes people to (1) make an unprofessional impression, (2) become totally incapacitated with fear when it's their turn to talk, or (3) completely refuse to participate in presentations at all. In short, public speaking scares the crap out of 99 percent of all people. Just ask Jerry Seinfeld, who put forth a hilarious un-joke that's paraphrased something like this:

> *"In a recent worldwide study of human fears, it turns out that public speaking is No. 1. Death is No. 2. Can you believe that? Death came in at No. 2. So, apparently when you go to a funeral, the average person would rather be dead in the box than have to stand up and deliver the eulogy."*

If you're good on stage, at home with a microphone, or simply comfortable at the front of the room, you need to make sure the company you work for sees you in action. Most rising stars are reasonably good public speakers, meaning they can refrain from visible shaking, keep a steady thought, stay on track with the flow of their presentation, and appear professional. But the elite of the MVPs are quite literally in the zone up there, whether speaking on behalf of themselves, their department, the entire company, or the industry as a whole. When all eyes have to be on someone who can handle the pressure and say the right thing, the elite speakers perform very well and "make the play" with their candor and convincing words. Picture Wayne Gretzky whispering to his teammates during the face-off huddle with eleven seconds left on the clock: *"Feed me the puck."* He knows he's the go-to-guy and so does everyone else.

Excellent public speakers are engaging, improvisational, conversational, and persuasive, and they make it look easy. If this sounds like you—high five—your talent for public speaking will be one of the most valuable bonus qualifications you could possibly have.

No question about it.

A talent for public speaking supports and strengthens all other bonus qualifications. An ability to articulate your thoughts intelligently and persuasively to others, no matter how many people are in the room—be it one or a 100 or a 1,000—will ensure that you are seen and heard and likely received well by others. Think about that. How can you accomplish anything in your career if you're not seen and heard and received well?

Lastly, there's another reason that being an excellent public speaker is incredibly valuable. The laws of supply and demand will be forever in your favor because 99 percent of all people completely suck at it.

THE CRITICAL PATH FROM GOOD TO GREAT

Your answers to the *"What else you got?"* question will dramatically affect the trajectory of your construction career. Your basic qualifications are essential and they'll keep you in the game, but your bonus qualifications will bring you far more opportunities, much more quickly. You need more than the basics to get ahead, stay ahead, and earn your way to the top. Whether you're a recent college graduate with your construction management degree in hand, a mid-level project engineer, estimator, seasoned project manager or superintendent, project executive, or vice president of operations—your basic qualifications may get you an interview, but your answers to the *"What else you got?"* question will get you the job, the raise, the promotion, the big bonus, and anything else you want along the way. Your growth in both opportunities and income will follow in direct proportion to the growth of your bonus qualifications and your ability to articulate their value.

It is critical to live in this awareness and work diligently to enhance your differentiating strengths continuously over the life of your career. All the while, remain on constant alert for opportunities to deploy your unique talents and relationships for the benefit of the entire team. Doing so catapults you to MVP status and establishes you as a leader, because keeping your ROI on graceful display sets an excellent example for others to follow.

Know first for yourself, and then teach others, that the critical path from good to great in the commercial construction industry no longer runs solely through construction.

It's essential but no longer enough. The industry is well past its competitive tipping point. Companies and professionals who fail to learn this stay good. Something *else* is now required to be great.

Figure out what else you got.

Sell it to the highest bidder.

Then help your company do the same.

Ignore your LinkedIn profile, resume, or bio for the moment. Ignore what you're going to say at your next interview or discussion with your boss about your future. Don't worry about all of that for a few weeks and spend some time in honest reflection of who you really are as a person who happens to work in construction.

What are the skills, talents, experiences, and relationships that define you? Focus on them, looking for the ones that have a direct value proposition to the commercial construction industry at large. Ignore the basic qualifications that all legitimate construction professionals have, unless you are highlighting a plausible and provable spin on how you get better results than the next guy. You're looking for your own unique set of bonus qualifications, so you can enter them into a chart similar to the one below.

If they're slow to come to you, seek input from a trusted colleague who has seen you in action or close friends and family members who will be honest with you. Ask for their professional opinions on what bonus qualifications they see in you. Chances are they'll find at least one that you were not aware of all by yourself. It can be easy to overlook something that you do naturally very well, until others point it out to you and confess that they cannot do it at all. You may be pleasantly surprised to learn that you have more bonus qualifications than you had previously thought; however, don't get discouraged if you're running short. Just get busy working on the one you would be most proud and advantaged to have.

Build out this chart over time. It's your cheat sheet. Center your personal dog and pony show around everything in it. As you learn to tell powerful stories that illustrate why each bonus qualification has a proven return on investment, you'll start cashing them in like a check.

MY BONUS QUALIFICATIONS	VALUE PROPOSITION & RESULTING BENEFITS
1. I have a great relationship with Frank Facilities.	a) My company gets work from Healthy Hospital in large part because of my relationship with Frank. b) Healthy Hospital has a lot of work coming up in their master plan.
2. I can train other people.	a) I'm a willing and capable teacher. When I learn something, I have a natural desire to teach others. b) I'm good at training. I teach in a way that fosters real learning and retention. c) Investment in me increases the ROI on others because I help raise everyone else's game.
3. I'm a compelling public speaker.	a) I'm very effective in sales presentations, helping to win projects. b) I speak often at industry events, which is great exposure and public relations for my company. c) I inspire and help others to improve their speaking skills.

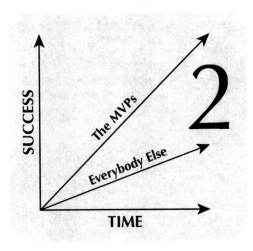

THE TRAILER AIN'T
THE FIELD

There is no substitute for real field experience in the commercial construction industry. Period. It's simply that important, and you can't skirt your way around it doing paperwork in the trailer. The trailer doesn't count.

FIELD EXPERIENCE MEANS FIELD EXPERIENCE

Let's start by getting clear on what field experience actually means to most mid-to-large general contracting and construction management firms. Field experience means *field* experience—not trailer experience. So that excludes most project engineers and project managers and their assistants, as well as project co-ordinators, accountants, administrative assistants, or any other member of a project team that works from a desk in the trailer on a jobsite. All of those positions give you valuable experience for your career—no doubt—they just don't give you actual field experience. There's a powerful distinction.

Field experience happens outside the trailer, in constant interaction with the foremen and tradesmen and other construction professionals who are actually building the building. To use specific titles, field experience is typically gained as a field engineer, assistant superintendent, project superintendent, or something equivalent. These are the positions where you are directly responsible for daily construction activities, quality control, scheduling, inspections, safety, and, above all else, subcontractor coordination—the most critical of all field responsibilities and the essence of construction management at its most basic level. That's why virtually all MVPs in the commercial construction industry have actual field experience on at least two or three sizable projects. It exposes you to, and makes you responsible for, the thing that contractors actually sell.

If your onsite experience is limited to the trailer, or you have yet to make it out of the main office entirely, you may be missing some of the credibility you need to overcome one of the most common objections raised against many construction professionals—*lack of actual field experience*—which is a fair objection, as most construction management professionals don't have it. It's advantageous to eliminate that objection by proactively seeking field assignments and sticking with them long enough to reap the rewards that follow.

No matter what anyone tries to tell or sell you, there are things best learned—and sometimes only learned—in the field. If you don't have any real field experience, find a way to get it.

As you make your way, here's a sneak peek into some of what the MVPs take away from their time in the field and the wisdom they look for when hiring their own project team members.

ALL SUBCONTRACTORS NEED TO MAKE MONEY

The single smartest thing you learn to do in the field is help your subcontractors make some money. The operative word in that sentence was *some*. Exactly how much profit subcontractors should make on any given project is another discussion entirely, but the primary point that everyone needs to grasp is this: You do NOT want your subcontractors to lose money. Not even one— because the failure of a single subcontractor can, and often does, cause a project to unravel.

Does that sound obvious? You would think so. But apparently it's not, because there are far too many companies out there that allow, and some that even promote, the exact opposite objective. As you go through your career, you'll run into companies that operate as if they're literally trying to drive their subcontractors out of business, with little appreciation as to (1) why that is un- ethical and (2) why it's not productive. Stay clear of them. If you find yourself working for that kind of company right now, get out of there before they brainwash you into thinking it is normal or justified to make a buck—that's total B.S.

Construction professionals on their way to the superstar ranks possess the uncommonly common sense that all project stake-holders need to operate in the black in order for the project to be successful. And since subcontractors collectively hold north of 80 percent of the stake in most nonresidential construction projects, it's imperative that each one of them make some money.

The MVPs in our industry know this is critical, and they act accordingly. They bring a servant-style leadership to the jobsite, which means they feel it's their responsibility to *serve* the sub-contractors. They work hard to inspire full cooperation among all subcontractors by delivering full cooperation at the management level. For example, a great project superintendent does every-thing in his power to make sure that his subcontractors have what they need to succeed. Here are some examples:

- He removes physical and managerial obstacles.

- He helps them prevent and solve problems.

- He gets prompt answers to their questions.

- He strives to maximize their productivity so cash can flow as quickly as possible.

And lastly, no project superintendent in his right mind ever holds money back from a subcontractor who has legitimately earned it.

Reputable general contractors know this is the best path to the desired win-win over the long haul. Generally speaking, when the subcontractors are making some money, the general contractor has a shot to make some money, too. When even one subcontractor is losing money on the job, the entire team is affected, and everyone's chances to make the desired profit, or any portion thereof, begin to drop. In fact, if the subcontractor in trouble is particularly integral to the success of the project, everyone's chances for a profitable job drop dramatically. Why? Because when a subcontractor starts to lose serious money on a project and they can't see an achievable path to profitability, they often surrender to the impending doom and begin writing it off in their mind, and their books, as a lost cause. This brings the entire project's productivity to a screeching halt. If you watch closely, you can see a subcontractor's telltale signs when this is happening:

- They start missing coordination and safety meetings.

- They fall behind on routine paperwork.

- They inflate the cost of pending change orders in hopes of making up some of their losses.

- They lay false blame for their lack of progress by saying that they cannot do X because of Y, and Y is not a logical or valid reason.

- They pull manpower from your (losing) project and send them to another project where they still have a chance of making money.

- They fall behind in their work, delay the other subs, and thus delay the entire project.

Once a subcontractor is that far gone it's hard to bring them back and very costly in one way or another, so the MVPs strive to prevent it from happening in the first place. Even if the sub-contractor's contract was a bad one from the start—because they took the job too low, or had a bust in their cost estimate, or they lost their foreman, or whatever—the MVPs are empathetic to their subcontractors and work hard to help them grind out the best possible outcome. This servant-style leadership and ability to pro-mote the individual success of each player in order to protect the success of the entire team is a critical leadership skill that is developed phenomenally well in the field.

You might not like a certain subcontractor. You might never choose to work with them again on any future projects—if you can help it. You might even think they are completely unethical, unprofessional, unqualified, incapable, or any number of other unfortunate things. However, if they're currently on your project and the options for replacing them are unacceptable, your best bet is to get them to the finish line, even if you have to carry them. The success of the project is dependent upon the success of the subcontractors, whether you selected them or not. If you want to make a buck, help them make a buck first.

THE ALMIGHTY F WORD

The F word we all know and secretly love is pervasive in the field. It's used at least once in nearly every sentence of any conversation, whether heated or friendly. It's also grammatically versatile as everyone's favorite four-letter noun, adjective, verb, or exclamation. But there's a five-letter F word that's far more important out there. It's arguably the most powerful word in the field—bar none—because without it pretty much nothing gets done. The almighty F word is *favor*.

In the world of general contracting and subcontracting, when someone does something for someone else that he was not obligated to do in his contract, that's a favor. Or when someone does something for someone else that he perceives to be above and beyond his own definition of reasonable or customary practices, that's a favor, too. It happens all the time to keep the job moving and the money flowing. To paint the picture further, when a subcontractor does a favor for another subcontractor, they're actually doing a favor for the general contractor as well, because there are no direct contractual relationships between the subcontractors. They don't technically owe each other anything, so when two subcontractors work something out between themselves harmoniously, through whatever means, the overall project benefits. Respectful general contractors acknowledge this cooperation, especially when they were the ones who caused the problem in the first place, which is often the case.

Favors are granted and received subjectively because perception rules. Serious arguments are downright cliché when it comes to defining whether or not something is obligated under contract or being done as a favor. Regardless of what's written and signed in a drawer somewhere, contractors keep their own score of who owes whom and why. It's all about how the facts of any given situation are interpreted and what is argued to be fair, which is ironically another four-letter F word. Except that one is rare. You could easily drop dead trying to find *fair* in the field. It might not even exist. Someone is always compromised, or perceived to be.

The MVPs know this is part of the game. You have to keep some sort of score and attempt to make things right when necessary—or suffer the consequences. Hell hath no fury like a righteous contractor who thinks he's due. It can get ugly.

Giving favors, receiving favors, and acknowledging of favors is a dance you learn in the field. Regardless of the scope of services outlined in a contractor's contract, there is a general understanding of who typically does what on a commercial construction jobsite. Somehow everyone is assumed to know what that is and act accordingly, or expectations are violated—which is a fancy way of saying that people get pissed. With a few exceptions here and there, most subcontractors come to a project knowing what they're required to do (and not do) without referencing their contract. They've simply learned over the years what is typically expected of their trade versus other trades, and, similarly, they've learned what is expected of the general contractor, too. So, regardless of the paperwork, there's a customary set of expectations between subcontractors and general contractors that everybody comes to understand through some sort of osmosis, and then another equally important set of expectations among all of the subcontractors themselves. Unless something to the contrary is brought to everyone's attention and everyone agrees to it—which is a tall order—everyone will rely heavily on what they typically expect of one another in this unwritten code of jobsite justice.

It's imperative to learn this code, and you cannot learn it from the trailer. It gives you a frame of reference and helps tremendously to explain why it's so common to see a foreman or tradesman staring someone down with angry eyes yelling, *"Hey!*

What are you doing?" Clearly someone's expectations are being violated, and, in order for everyone to get back in line, a favor is about to be exchanged and mentally noted. It could be as minor as someone needing to stop what he's working on, come down off of a ladder, and move his job box over twenty feet. It doesn't matter. It gets noted. It could be a coordination bust where three subcontractors find themselves trying to work in the same little room at the same time. Or one subcontractor might accidentally damage another subcontractor's materials or equipment. Or the staging area needs to be moved for some good or bad reason. No matter what the cause, the effect is the same. At least one subcontractor will be thrown out of step in the dance of expectations, and fingers will start pointing back and forth at one another until everyone is forced to iron out the favors required to get everyone back in line.

Sometimes change orders or back charges will be warranted to right the wrongs along the way; however, in the big scheme of things, the number actually issued is probably a drop in the bucket compared to the number avoided.

> *People who haven't spent enough time in the field would be shocked to know how much construction activity actually goes down on favors, and how many change orders and back charges are avoided every day by the almighty F word.*

This form of cooperation is a truly remarkable aspect of the commercial construction industry and you cannot learn to appreciate it anywhere else but in the field.

Not in the trailer.

The field.

THE ONE CALL YOU NEVER WANT TO MAKE

You have to prevent people from getting hurt on your job-site. End of story. That's not a politically correct stump speech. It's the real deal. Your motivation to keep people safe can be a legal one, stemming from the (huge!) underlying corporate risks and liabilities rendered by the law—but in case you need personal motivation to actually do it, here you go:

> *When someone gets seriously injured or killed on your jobsite, it changes who you are. It never leaves you.*

So you must do everything in your power to promote and enforce jobsite safety regulations on a daily basis, and then hope and pray that your efforts are enough. Not because it's the law, because it's right. And your integrity is on the line.

If you do everything that you can to keep everyone as safe as possible, and someone still has an accident that results in serious injury or death on your jobsite, you'll be able to live with it. But if you fail to do everything possible to prevent that accident, you'll carry some of that guilt on your shoulders for the rest of your life.

Contractors are quick to complain about the cost of work-man's compensation insurance, but slow to acknowledge why it's so incredibly out of control. Yes, much of the exorbitant costs are caused by lawsuits, politics, and insurance companies, but the rest of it is caused by the undisputed fact that being a construction worker is a very, very dangerous occupation. If you don't believe that, ask some guys who've been around for a while and they'll tell you some devastating stories.

It's easy to forget all of this until you're dialing 9-1-1 from your cell phone, leaning over a little girl's father who is fighting for his life after falling twenty feet from an extension ladder onto two pieces of exposed rebar that have impaled his kidney, spleen, and groin. Or watching a roofer lose his footing, and, while frantically grasping to regain his balance, he accidentally tips a nearby bucket of 400-degree tar down the entire right side of his body. Or you experience the absolute horror of learning that one of your workers was just decapitated in an elevator pit when the temporary elevator operator made a deadly mistake.

Or you get hurt—which also happens.

Whether you're calling 9-1-1 to help someone, or someone is calling 9-1-1 to help you, you cannot appreciate the danger of the field until you live in it ten hours a day. The MVPs understand this intuitively as a moral imperative, but also deductively in their constant quest to identify and manage risks. Jobsite safety is the riskiest of all risks and it's managed in the field.

Not the trailer.

The field.

CONTRACTORS SELL VERBS

Here's something most construction management schools don't teach you about the field, but they should. Project managers don't run construction projects—and project executives don't either. Not really. Not like everyone is led to believe they do, or the titles imply. Generally speaking, they control the documentation, administration, and the financial management of their projects—all of which are essential to the construction process and client relations. They're also certainly held accountable for the success of their projects. For the most part, however, they don't run the *work*, so they don't run the job. The guys who run the field run the job. Project superintendents and their assistants, working in cooperation with all subcontractor foremen, are the people who actually get the building built. They direct the labor, which results in work-in-place, which results in payment.

The term "project manager" can be deceiving because it implies the highest level of authority on a project, which may even be technically true on a piece of paper somewhere. Yet on large, complex commercial construction projects, the superintendence team will often carry more influence over the speed and quality of actual work-in-place, regardless of the organizational chart. Our industry's MVPs learn this quickly, particularly in sizable companies. They discern and appreciate the difference between driving maximum work-in-place every month and everything else that exists to enable it and support it. Again, there's a powerful distinction.

There are pointless arguments about who outranks whom and who's more important than whom in the chain of command on a jobsite, ad nauseam. No matter where you stand—keep your ego in check—everyone is obviously incredibly important. Go ahead and argue any way you want, if you must, but the MVPs know:

What we do for a living is drive work-in-place. That's what we get paid for and why we get paid at all. Everything else is supportive by definition.

This is one of the key insights or aha moments you uncover in the field, if you stay out there long enough and pay close enough attention to see what's really going on. Then it sinks in and stays with you as you climb the corporate ladder, influencing the way you see the industry in a really smart way.

Some of the MVPs in our industry were once world-class project superintendents themselves—magicians in the field and masters at driving work-in-place. Fortunately, some are still out there and have no desire to do anything else. A few of them defied the odds and never worked in the field at all; nevertheless, they all know this truth deep down in their bones and make decisions accordingly:

Since work-in-place is WHY everyone gets paid, the guys who drive it hold the cards and run the show. They do the thing we actually sell. They don't enable it or support it; they DO it.

Many construction management professionals fail to grasp this, or maybe don't want to, and it gives the MVPs an edge. They maintain a respect level for high-performance field personnel that borders on reverence—and rightfully so—because guys who can drive work-in-place are money in the bank. And when they are surrounded by high-performance project managers, engineers, and assistants, truly amazing things can happen in the field. Projects with seemingly insurmountable odds are overcome every day.

> *The guys who run the field can be awe-inspiring. The MVPs pay top dollar for them, win work because of them, listen to them, and protect them at virtually any cost.*

If you ever get a chance to obtain field experience working for a general contracting or construction management firm that employs world-class project superintendents who maintain the highest level of authority on the jobsite—take it. Working under the wing of a seasoned leader who demonstrates a mastery of field coordination is the best possible way to learn two incredibly important things for your construction career: (1) how to actually build a building and (2) how to straight-up make things happen. Believe it or not, the latter is more important—because when you leave the field and pursue success at the executive management level, you'll need to know how to lead people and make all kinds of things happen.

Virtually all MVPs will agree that it's imperative for commercial construction professionals to learn the business from the ground up, which can only be done out in the field, where all of the action takes place. Everything else is at least one step removed from the service contractors actually sell. Contractors don't sell cost estimates. Contractors don't sell RFIs, shop drawings, or a CPM schedule. Contractors don't sell a change order log, an accounting ledger, or a punch-list. They don't even sell emails or reports. Contractors clearly need and use all of those things—as they're critical to the construction management process—but they don't sell them.

Contractors don't sell the nouns.

They sell the verbs.

Contractors sell the construction of a building: the actual *moving* of dirt, *pouring* of concrete, *erection* of steel, *pulling* of wire, *plumbing* of pipe, *hanging* of drywall, and *pounding* of nails. Sometimes contractors even sell the *making* of something incredibly beautiful.

But all of it happens in one singular place.

The field.

DO IT ANYWAY

Very few people will admit this out loud, but, in the privacy of your own head, are you intimidated by the idea of running work in the field? If so, don't worry. You're normal. It's really hard. Anyone who downplays this truth is throwing you false bravado or speaking from the ignorance of never having actually done it on a highly complex project.

The daunting responsibility of coordinating a group of contractors to drive work-in-place—particularly under ridiculous schedule and budget constraints—while ensuring everyone's physical safety and strict adherence to the approved plans and specifications is difficult to fully appreciate until you've experienced it for yourself. But it's much easier to face the job when you realize that everyone experiences various levels of intimidation and anxiety out there. Everyone is facing the same endless array of unanswered questions, unending problems, and unrealistic expectations. On top of that, everyone is doing his best to get along with a wide variety of complex personalities and egos to boot. Why do you think the emotions run so high on jobsites? Every single person out there is literally trying to survive it, make a living, and walk away with some pride in his work. Most people just don't admit how hard it really is. So go do it anyway. It'll put some serious hair on your chest and give you the confidence to know that you can do anything else after that.

Ultimately, the MVPs know the field is a test of actual construction knowledge, leadership skills, and courage. It's just as much about proving to *yourself* that you can excel in the field as it is about proving it to anyone else. It's the real deal out there. No matter what happens in your career after the field, you'll end up pretty darned fearless. And you can take that with you wherever you go, in this industry or any other.

Regardless of your current position, there's a great way to begin accessing some of the wisdom found in the field. It's time to cherry pick two very specific guys and add them to your circle of mentors.

Here's the first one. Think for a minute about some of the best project superintendents you know. We're talking about the dying breed—the old school, grey-haired guys who love it out there and actually know how to run a job. Of all those guys, who is the best of the best? Write down his name and skip to the next paragraph. If you don't have the privilege of knowing any great superintendents, ask someone you trust for advice and get an introduction.

This guy is your next mentor. If you're relatively young in your career this will feel natural. If you've been around for a while it may feel less natural—but it doesn't matter how old you are or what your title is on some organizational chart, just do it. Spend some time with him because it's very likely that he's already forgotten more than you'll ever know about how to actually build a building. Lest we forget, that's the business we're all in.

Reach out and tell him straight up why you're calling. Ask if you can meet sometime in the next month, after hours at his jobsite or somewhere to grab a bite to eat or whatever. If he's not local to you, arrange a time to speak on the phone, but do it in person if at all possible. If you're struggling for the right words to jump-start the conversation just say, "*I need to learn more about the field and you are the best superintendent I know.*"

It may be awkward at first, but if you're sincere, he will sense it and come around. Look, stereotypically speaking, there are three things you can take to the bank when it comes to great project superintendents and their personalities. First, they're hilarious. It may be hidden under a tough and serious exterior, but they usually have a great sense of humor that comes out eventually because everyone needs a few good laughs to stay sane out there. Second, they're Academy Award-winning storytellers about their projects and the people they've worked with along the way. They love to talk about the good, the bad, and the frigging outrageous experiences they've had, especially if they sense you really want to learn and know what it's like to walk in their boots. And third, they're grossly under-appreciated and under-tapped for their professional input, so know that your request is a compliment they well deserve and your admiration will be appreciated.

When you meet, get him talking and listen for the wisdom to drip out slowly. Chances are he'll soon have you trying to drink from a fire hose. Here are five good questions you can ask him to prime the pump:

1. What do you wish everyone in the office really knew about the field?

2. What is the single most important thing required to keep a project on schedule?

3. What separates a great sub from a good sub?

4. Who's the best project manager you've ever worked with, and what did he do that made your job easier?

5. Who taught you? Who's the best superintendent you know and why does he rank No. 1 in your mind?

Here's the second one. Think of some of the best subcontractor foremen you know. Focus in on concrete, mechanical, electrical, or plumbing foremen, because those four trades are typically on the job the longest, so they interact with the most work-in-place over time. Again, go for the dying breed—the guys who are true experts in their trades. Of all those guys, who is the best of the best? Even better, pick one from each of the four specialties.

Repeat the same mentoring process noted above, but this time hone in on questions like these, where you'll gain valuable insights you can share with your colleagues to help everyone in your company:

1. What does it mean for a general contractor to actually coordinate a job properly? What should happen at coordination meetings? What should be on the agenda and how should they be conducted?

2. With respect to your trade, what are some of the most costly mistakes general contractors make during the first thirty days of the project? Midway in the project? Near project completion?

3. What are the most common areas of conflict you encounter with other trades? Are there ways to prevent

this conflict that most general contractors don't know about, or fail to implement?

4. Other than labor, what are some of the most expensive line items in your world? Ask him to finish this sentence: "You'd be shocked to know how much we have to pay for _____."

5. What's the most important thing you could teach me about (his trade) that 99 percent of all architects and general contractors don't know anything about, but they should because it causes all kinds of problems.

Stay in touch with everyone and take the mentoring process seriously. It's priceless for your career. These guys are wicked smart and they've seen it all. No one in a suit and tie will ever teach you more about how things really work.

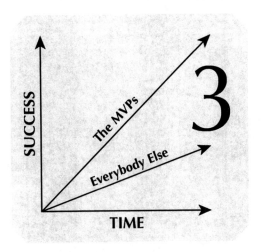

GET YOUR OWN
JERRY MAGUIRE

Have you seen the movie *Jerry Maguire*? If not, watch it. If you have, watch it again, because this time you're going to watch it from an entirely different point of view. Ignore the romantic comedy plot lines like *"You had me at hello"* and *"You complete me"* and focus in on the incredibly illuminating story of the athlete-agent relationship in professional sports. It's powerful and amazingly applicable to your career. If you want to join the MVPs in the commercial construction industry, you might want to buddy up with your own Jerry Maguire.

He can help.

In more ways than one.

A SIMPLE ALIGNMENT OF INTERESTS

Virtually every professional athlete, musician, actor, or author has an agent and many other occupations do as well. The agent represents these professionals in their respective marketplaces, and they help procure and negotiate employment contracts and endorsement deals for their clients. In addition to their powers of persuasion and negotiation skills, agents also bring valuable industry relationships to the table, which strengthens their value proposition. In short, excellent agents have excellent connections. They know people. They open doors. And when they're properly motivated—as in the notoriously famous line from the movie—they can really *"show you the money!"*

It all stems from a simple alignment of interests. Generally speaking, agents are compensated as a percentage of their clients' income from employment contracts and endorsement deals. So obviously what's good for the agent is good for his client. When the agent helps secure his client more opportunities and money, the agent typically gets more opportunities and money right along with him. It's a true win-win relationship.

In the commercial construction industry, the closest equivalent to an agent is an executive search professional or recruiter. These guys are the potential "Jerry Maguires" of our world. Not just any old headhunter, but rather true executive search professionals who specialize in the U.S. commercial construction industry and do their job well. The ones who've built trusting relationships with many construction companies (a.k.a. the teams) and many construction professionals (a.k.a. the players) and are looking to connect the best with the best from both sides, for the right positions, and make a buck in the middle. That's pretty much their gig. It's not a perfect analogy, but if you tweak it slightly it works. More importantly, it works for you, and the company you work for, if you work it right.

THE MVPs HAVE AGENTS

Here it is, plain and simple. Superstars have agents. The industry you work in doesn't matter. So if you're a superstar in the commercial construction industry you should have your own Jerry Maguire, too—someone who's essentially pulling for you, deal-making on your behalf, and just plain rooting for you out there. And keep in mind, there are superstars at every level. There are rookie superstars with tons of potential, mid-career superstars, veteran players, and high-level executive management superstars. No matter where you are in your career, if you're really good, you can command the attention of the top executive recruiters who'll be eager to develop a win-win relationship with you.

A little context helps.

It works like this.

BOTH SIDES OF THE FENCE

There are three things that count in the world of executive search professionals: relationships, relationships, relationships. Top-notch executive recruiters work very hard to develop long-term, trusting relationships with people on both sides of their matchmaking fence, which are (1) the companies they recruit for and (2) their network of construction professionals, from which they pull candidates to fill positions or seek referrals to other candidates. They have to do this in order to survive and develop a good reputation.

Now think money. Executive recruiters in the commercial construction industry are generally paid directly by the companies they recruit for, and the payment is typically either a flat fee negotiated up front or a percentage of the compensation package offered to the successful candidate. So technically speaking, they deem their *official* clients to be the companies they recruit for, because that's who signs their checks; however, in a very important and real sense, they also consider the commercial construction professionals they place with those companies to be their clients as well. They may be paid by one master, but they serve two masters, because they cannot effectively serve one without the other. The most-respected recruiters know this intuitively. The ones who don't are bridge-burners and they make more enemies than friends. Who pays who in the commercial construction industry is different from the sports agent analogy, but the mutual best interests of all parties remain aligned, and, therefore, the analogy holds. What's good for you is still good for your executive recruiter and vice versa.

Top recruiters consider themselves to be an extension of their clients' human resource department, with a fine line of separation. They care about the companies they recruit for and seek to earn their repeat business by helping them attract the best possible talent available in the marketplace for the opportunity and compensation package being offered. Similarly, recruiters' passion for good matchmaking extends its reach over the fence to the growing network of construction professionals they've come to know over their careers. Their network is their lifeline, their connectivity to the industry, their direct source of potential candidates or referrals to other possible candidates. So the good guys work equally hard to earn the respect and trust needed from their network in order to be successful in the placement process, resulting in satisfaction on both sides of the client fence.

JERRY MAGUIRE IS RARE BUT THERE

If you ask around, you'll likely hear a mixed bag of emotionally charged opinions about what it's like to work with executive recruiters. Some construction professionals swear by it and some curse it. The ones who curse it do so loudly and emphatically, with all kinds of horror stories about how they were screwed over by so-and-so when they went through a headhunter to get hired or hire someone else. Many of those horror stories are true. There are clearly unethical recruiters who will recruit professionals into your company, and then turn right around and recruit others away from your company. There are also ones who are only in it for a quick buck. Those guys will woo someone into a position that really isn't a good fit for either party, but it results in a nice fee, so they take it and run.

It happens.

When you see that kind of behavior, avoid that person and move on, but don't write off the value of the entire service just because some executive search professionals—or even many of them—do it unethically. The commercial construction industry has no room to talk anyway. Contractors have a collective reputation that hovers one step above politicians and one step below used-car salesmen. Is that reputation fair? Largely not, but it probably depends on whom you ask. There are bad apples in every bunch of every industry. That just makes the really good apples even more valuable.

WELCOME TO THE CLUB

Although the MVPs rarely discuss it openly, having your own Jerry Maguire is one of the best-kept secrets to success. Relationships within the executive search community can feel like membership in a quasi-private club, reserved for the few who understand how it works and how to maximize the benefits. Here's a glimpse into some of what the insiders know.

Insider Tip 1:
Jerry Maguire can help you find a job—duh.

The most obvious and personal reason for having your own Jerry Maguire will be his ability to help you find a new job when you're looking for one. He's helpful because he'll have advance knowledge of (1) which companies within his client base are actively hiring, (2) which companies outside of his client base are actively hiring, because they often work cooperatively with other recruiters and keep a keen eye on the marketplace, and (3) what specific positions all of the hiring companies are currently trying to fill. Advance knowledge means that Jerry Maguire is likely to know what's coming around the corner in terms of potential opportunities before they reach the construction community at large, or any employment or social networking websites. If you're interested in any of the positions he's trying to fill, his connection will be helpful, but most importantly, his influence to recommend you will be the real key. You'll see why at the end of the chapter.

Insider Tip 2:
It doesn't matter where Jerry Maguire hangs his hat.

Just like the end of the movie implies, it's your relationship with Jerry Maguire (the person) that counts, not your relationship with the company he happens to work for. If your Jerry Maguire leaves his current recruiting firm and goes to work for another company, or starts his own firm, he'll likely take the majority of his client relationships with him. Or at a minimum, he'll retain some level of influence with those companies. So it's in your best interest to remain in close contact with your Jerry Maguire, no matter where he hangs his hat.

A true professional will be forthright if he makes any career moves that could affect his ability to serve you. He'll likely refer you to someone else who can help you more effectively, if necessary.

Insider Tip 3:
Jerry Maguire knows the street and the grass.

The top executive search professionals in our business know the street, meaning they have a pretty good feel for what's going on in the industry as it relates to human capital, such as:

■ What companies are growing and hiring

- What companies are shrinking and firing

- Current salary and compensation packages

This is particularly true within their area of expertise or geographic niche. For example, some of our industry's search professionals specialize in placements for large-scale general contractors in, say, the Pacific Northwest and will therefore have a wealth of information about the companies in that part of the country. Another firm may specialize in small-to-mid size mechanical and electrical contractors across the entire United States. Some focus on placing construction managers with substantial experience in certain types of projects (i.e., healthcare, airports, or correctional facilities). Careful research will reveal which firms have established favorable reputations and market mastery in the areas of specialization or geographic locations that are most relevant to your career goals.

Maintaining a constant feel for this type of information is a full-time job, and it's one of the ways that executive search professionals earn their living. Therefore, having your own Jerry Maguire is a great reality check. He can give you bits and pieces of current information from the street when you need it.

In addition to the street, Jerry Maguire often knows about the grass. For example, some people reach out to their Jerry Maguire to discuss the general state of their career, their goals, and any potential opportunities that may be inside or outside of their current company. These kinds of confidential discussions can be helpful and enlightening, because he may have some insights about whether or not the grass is greener somewhere else. It may be. It may not be. Either way, the MVPs often tap into their Jerry

Maguire for advice before making any major decisions, as it can be a much more discreet way of gathering some of the information you need.

Not only will his knowledge of the street and the grass help you personally in your own career, it will also help you protect the best interests of the company you work for when building your own internal teams, which plays nicely into the next point.

Insider Tip 4:
Jerry Maguire can build you a superstar team.

As your career takes you into middle- and upper-level management positions, having your own Jerry Maguire will take on another equally important role for you. Whether or not he recruited you, he can now recruit *for* you and help you build a powerhouse team of construction professionals. For example, if you're a vice president of operations, he can help you recruit appropriately experienced superintendents and project managers and other team members when you need them. If you're a chief estimator, he can help you recruit excellent estimators. If you're a senior vice president of business development, he can help you recruit candidates with influential relationships and a track record for revenue generation. And so on and so on.

Why might you need this help? Some construction companies are able to successfully fill open positions from within their current organizational structure, by moving people laterally or promoting people up through the ranks. This is often the best solution and when it works it works well. But there are also situations when it's not ideal or even possible to fill an open position from within and you'll need to recruit someone from outside the company to join your team.

In some cases, you'll be able to handle this hiring process on your own, or through your company's human resources department, but a growing number of executive managers in the construction industry are relying more and more heavily on the expertise, relationships, and professional recommendations of their own Jerry Maguire to help attract and retain the best talent. So it's a good idea to have yours locked and loaded.

Insider Tip 5:
Jerry Maguire is free P.R.

Last but not least, having your own Jerry Maguire is like having another kind of agent working for you for free—your own public relations agent—and you can't beat it with a stick. This alone is reason enough to have him in your pocket.

Keep an open mind and go back to the sports analogy again. Say you are an NFL receiver or a MLB pitcher or an NHL goalie, and your Jerry Maguire thinks you're the best he's ever personally seen play the game. At that point, he's not only your agent; he's your fan, and his relationship with you makes him money. Being your agent is bragging rights for him. It helps him attract other clients and grow his network of contacts, and everything else. Use of the word *fan* may be a creative stretch but don't miss the point.

Now translate that into your construction career, whereby your Jerry Maguire deems you to be the best project executive or business development manager or vice president of preconstruction that he's ever met. When this happens, and it does, human nature takes over and works to your reputational advantage—for how do you think your Jerry Maguire will talk about you to his clients, colleagues, and other construction professionals?

The answer is very, very highly.

And word gets around.

THE MVPs GO IN THE FIRST ROUND

In summary, the real reason you should watch (or re-watch) *Jerry Maguire* isn't to remind you that all superstars have an agent. There's another more important reason. The true spirit of the movie takes you past the necessity of having an agent, into the titanic advantage of having an agent that sincerely believes in you.

Not simply an agent that will go to bat for you by putting in a few good words. Lots of people go to bat for other people, with little to no concern about the actual outcome. Remember from the tips above that we're talking about having a Jerry Maguire who's your fan. He thinks you're the best of the best when nobody's looking and nobody's asking. Someone that truly believes, in his gut, that you are the best person he knows for the job you want next. Therefore, you are always his first round draft pick (so to speak) to play your desired position for virtually any team. Whether it's spoken or unspoken, somehow between the two of you this is understood and, therefore, your name is always called out in the first round.

That's how the MVPs and their Jerry Maguires roll. When he believes in you, and you know it, the win-win relationship starts to work. The cynics will scream otherwise, but our industry does have honorable representatives out there who are trying to do the right thing for everyone involved, which is to facilitate the best possible match between the right player and the right team for the right position. That's Jerry Maguire's job.

If you want to get in on it, your job is to earn his admiration and be worthy of his first round draft pick—so he always recommends you above any of your competitors—on both sides of the hiring equation. That's exactly what the MVPs do and why they develop such a seemingly unfair advantage. But it's not unfair; they just have massive leverage. For example, the MVPs enjoy knowing that when they need to hire someone, their Jerry Maguire will sell them (and their company) hard to potential candidates by saying something like this: *"I know you have several options here, but I'm telling you, the guy you really want to work for is this guy right here. I've been doing this for twenty years and I can tell you he's the best. Everybody wants to work with him."* On the flip side, when the MVPs are looking to make a personal move, they know their Jerry Maguire will pitch them as the #1 candidate for whatever position they play or want to play—but it all starts with his sincere belief in you. There's no way around it. You have to get picked in the first round. That's how the game is played.

Sell yourself to Jerry Maguire first.

When he buys you'll have two people selling you.

Do you already have a relationship with an outstanding executive search professional that could be leveraged into your own Jerry Maguire? If so, skip to the next paragraph. If you need help finding one, ask someone in the industry that you trust for a referral or connect with Coty Fournier on LinkedIn.com and request a list of recommendations.

Take a cue from the MVPs and follow these seven guidelines to increase your chances of success as you press forward:

1. Prequalify him first. Make sure he can show you a proven track record for success in commercial construction placements. Anything he has done outside of the industry is completely irrelevant. Ask him for a list of his clients, and how many people he has placed at each one in the last five years. Look for him to drop a few MVP names to show you that he has the right relationships. If he holds back, he might be waiting to learn more about you first. But if he doesn't eventually drop a few power player names that you recognize, he probably doesn't have any to drop. Move on.

2. Don't wait to establish a relationship until you need a job, as that's usually too late. You can *want* a new job, but *needing* a new job is an entirely different psychological state and much harder to sell. Start now wherever you are.

3. Know who you are and what you want. It's very hard for executive search professionals to get excited about you when you don't know who you are and what you want. Be clear. Be very, very clear in conveying exactly what you're trying to achieve in your career, both short and long term.

4. Tell the truth. These guys need to know the truth about your employment history, salary requirements, personality, career goals, and more. They need to know what you're actually capable of doing and what you're expecting in return. Anything less is a waste of everyone's time and potentially damaging to both of you. Do not risk your reputation (or his) by stretching the truth or overtly lying about anything. Build a proper foundation for the relationship that will result in trust and respect so you have a real chance of helping one another over the years to come.

5. Be ready to sell yourself with rock solid answers to the *"What else you got?"* question highlighted in chapter one. They need to be front and center in all of your verbal storytelling and reflected in your

LinkedIn profile, resume, and bio. If you think about it from Jerry Maguire's point of view, it should be obvious as to why this is so important. Pretend you are in his position, and he's taking twenty calls a week from people like you. What can you tell him in the first ten minutes that'll grab his attention, impress him, and make him begin to believe that you might be more employable (or marketable) than the other twenty people he talked to that week? If you cannot sell yourself to him, he won't believe you can sell yourself to anyone else.

6. Give before you take. To kick things off, see if you can help him with a position he's having difficulty filling, or an introduction to a potential client, or whatever is on his mental plate. Maybe you can refer a friend that will help him close a deal or land a retainer. If you want him to realize quickly that you're a rising star or a respected veteran, then prove it by showing him that you know people, too.

7. Stay in touch. You don't need to be a hound dog, but every four to six months it's good to drop him a line, update him on your accomplishments, and see if you can help him with anything. Remember, he's taking twenty calls a week from people just like you, so it pays to stay top of mind.

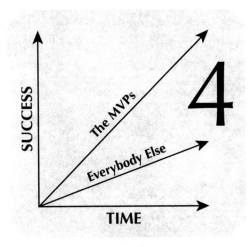

BEWARE OF THE PROJECT MANAGER MYTH

If you want something different than most people, you have to do something different than most people. So unless it's your ultimate career goal to become a great project manager—meaning that's your desired end game—beware of the project manager myth. It's the common notion that excellence in project management skills is the key to advancement in the commercial construction industry.

It's not true.

Not even close.

Few truly great project managers climb the corporate construction ladder much higher or succeed when they get there. Yet construction management schools teach the subject more than any other and the vast majority of all young construction professionals aspire to see the title of *Project Manager* on their business cards.

If this is you, and you want to be a great project manager, then by all means go out there and have at it. They play a very important role on every project and they're an integral part of every construction company. Our industry relies heavily on them and if you're one of the best you'll be rewarded handsomely. Just know that it may be tougher than you think to rise above it—should you ever decide to do so. It's an eye-opening experience when you break it all down and see what it actually takes to become a great project manager, the price you may end up paying for it, and why the industry's MVPs often choose to leap frog right over it.

WHAT IT REALLY TAKES

To become a highly respected project manager in any sizable construction company, you'll have to earn your way to the opportunity and then prove yourself to stay there. On average, you're looking at anywhere from ten to twenty years to make that happen and really own the title. The lower end of the range represents smaller companies who do smaller, less challenging projects. The higher end represents larger companies who do larger, more complex projects. Beyond that range entirely reign the elite companies capable of building mega projects that can span three to six years and total hundreds of millions of dollars in construction volume. If you aspire to be the lead project manager on one of those sexy $300 million sports stadiums, $500 million dollar airport expansions, or $1 billion mixed-use developments, prepare to invest more than twenty years of your career to earn a shot at it, plus another ten to fifteen years to actually manage a couple of them. Ask those who've already done it and they'll tell you. Those impressive notches on your belt are worn with pride, but they're also hard earned in blood, sweat, tears, and—above all else—in time.

It's possible to get a crack at your first project management role sooner, working for very small contractors who do very small projects, like office interiors, retail shops, and restaurants. Some construction professionals prefer that route and choose to remain in those less complicated circles. Just keep in mind, however, that egos come super-sized in this industry and that's not the glory most professionals are typically seeking. Nor the pay scale.

THE BEATEN PATH

Your first ten years will be spent working on a handful of projects in a variety of entry level and assistant level positions, designed to help you learn the ropes without the full responsibilities reserved for project managers. Depending upon the size of the company you work for and the titles they utilize, your first few assignments will likely include:

- Field engineer
- Assistant project engineer or project engineer
- Assistant project manager

You might also do a short stint or two in the estimating department, or help out with purchasing or preconstruction activities in between your project assignments. That kind of thing is very common and it brings up another important point. When you're ready to be a project manager on a certain size or type of project, it doesn't mean your company will have one ready and waiting in the wings for you. The timing of incoming projects rarely lines up perfectly with your readiness or availability. Even when your company is awarded an appropriate project that is tagged for you, there may still be delays in funding, permitting, weather, or any other number of things. Chances are you'll wait for a while.

Tick.

Tock.

Tick.

Tock.

When your first shot comes along, it'll likely be on a relatively low risk project, for good reason. You have to build your way up from rookie to reliable status on perhaps three to five projects that (ideally) grow successively in size and complexity. If they all go well, meaning each client becomes a raving fan and acceptable profit margins are earned, you'll be deemed a reliable project manager. Again, depending upon the size and complexity of all the projects in your portfolio, it will have taken you somewhere around ten to twenty years to reach your goal and far beyond that if you're dancing in the mega projects arena.

Of course there are always exceptions. Some people get promoted more quickly than the average because they're exceptional and they deserve it. More often than not, however, people get promoted quicker than the average due to economic factors. When the market booms quickly, sometimes construction companies are forced to promote people into project management roles before they're truly ready, or hire those with less experience than would normally be required. Neither is good. When people are prematurely promoted, they often find themselves walking around in boots they're not ready to fill, with an artificially inflated perception of their street value. When the economy shifts into a new cycle and the supply of project managers exceeds the demand, you can safely assume those people will be vulnerable to dismissal.

IT'S ALL GOOD UNTIL YOU WANT SOMETHING MORE

When you reach sustained success as a project manager, the company you work for will value you in that position and you'll be an important part of their operational structure. They'll utilize your resume and experience to pursue new projects and then rely upon you to manage them successfully. This is particularly true if you have extensive experience in certain niche markets that may be in high demand, such as healthcare, biomedical laboratories, higher education, airports, or correctional facilities.

This describes the essence of the *contractor-PM relationship* and, generally speaking, it works well for both parties. The contractor you work for benefits from your applicable experience and reliability to manage projects, which makes them money. You retain gainful employment and a steady stream of projects to further develop your skills, which makes you money.

Ahh . . . all is well.

Until you want to excel beyond project management into the ranks of executive management and corporate leadership. In which case, the project manager myth reveals itself, and you discover that you're lacking some very important qualifications to reach your career goals and logistically challenged to build the relationships you'll need to change your trajectory.

A PROJECT IS NOT A COMPANY

Being a great project manager does not automatically qualify you for executive management or corporate leadership in a sizable construction company. That's the primary consequence of the project manager myth in a nutshell and it comes as a shock to most people. Although it's difficult for some to swallow, it's important to understand that project management skills don't translate very well into corporate leadership, primarily because a project is not a company. They're not even like a mini company. So it doesn't matter how big or important or profitable each one of your projects turns out to be, project management is still only one of the myriad activities and responsibilities required to build and maintain a successful commercial construction company.

That doesn't shed an unimportant light on project management—it's clearly very important—it's just that all of the other things that happen before, during, and after project management are also important. These include such things as:

■ Marketing, public relations, and community relations

■ Business development

■ Estimating, purchasing, and preconstruction

■ Construction operations

■ Administration

- Finance and accounting

- Information technology

- Human resources

- Safety, legal, and risk management

- Corporate leadership

In particular, activities that directly precede the necessity of project management help to put it into proper context. More simply said, the MVPs who speak from experience will tell you that the hardest part of building and maintaining a commercial construction company is finding and winning profitable work—not managing the actual work. There are (huge!) challenges inherent to the industry's competitive landscape as referenced in chapter one; therefore, the people who know how to navigate it and generate revenue are essential to any corporate structure.

They're also increasing in value at a higher rate relative to everyone else.

Hint. Hint.

PROJECT MANAGERS ARE PROJECT-FOCUSED PEOPLE

Being a project manager requires you to be singularly focused on your project, to the exclusion of almost everything else. By definition, they eat, breathe, and sleep their projects and rightfully so. As a result, great project managers become project-focused people. They are exceptionally detail-oriented and task-driven, which are powerful skills they develop from working down in the weeds, analyzing and organizing various kinds of project information on a daily basis—and those skills are valuable.

On the other hand, highly effective executive managers and corporate leaders are not project-focused people. They may have once served and excelled in project-focused roles, but at some point they made a conscious decision to become company-focused people and relationship-focused people. Corporate leaders possess an expanded set of skills relative to project managers, born from their natural talents, personal strengths, and well-rounded career choices . . . and the MVPs among them are big-picture thinkers and true visionaries.

The heart of the project manager myth is really all about perspective. It's about determining from what perspective you're most comfortable thinking and what things you most naturally think about. Detail-oriented and task-driven project managers tend to fly around 10,000 feet—high enough to see the entire project, but low enough to remain immersed in all the details. Within the boundaries of their project, they know exactly where all the pieces and parts are at all times, and this is a natural, comfortable perspective for them. Conversely, executive managers and corporate leaders fly at 30,000 feet. From this vantage point, they have their finger on the pulse of the entire company and all of its functions and needs. This is their comfort zone and the highest and best use of their skills. When they need details, they drop down to 10,000 feet to find and analyze them, but they quickly climb back up to 30,000 feet before they make any decisions based on that information. Their natural perspective is the state or direction of the whole—not the parts.

Effective leadership at 30,000 feet stems from a general competence in all departmental functions versus an expert level of knowledge in any one of them—combined with the relationships and skills required to ensure the company's long term success and profitability. It doesn't mean that seasoned project managers (and other project team members for that matter) cannot learn to adjust their perspectives and become more company-focused and relationship-focused people over time. They certainly can and some of them do; however, because the project manager myth is so prevalent, many project managers don't recognize the need to change their perspective or their lack of experience in other critical company functions.

Dr. Phil may be over-quoted, but he's right on this one.

"You cannot change what you don't acknowledge."

OUT OF SIGHT

Warning! This one's going to hurt. Largely because it's not fair, but it's true, so it's important to be aware of it as you make decisions about your career path.

The commercial construction industry suffers from an inherent separation between the people who work in the field and the people who work in the main office. This physical separation creates challenges for everyone involved, as high-quality relationships are difficult to develop and maintain outside of close proximity. With few exceptions, the executive management team and the company owners go to work every day at the main office, while the project managers go to work every day in the field.

These logistics produce a completely undeserved disadvantage for project managers, but a disadvantage nonetheless. It's no one's fault. It's just the way it is. Project managers are out of sight and therefore out of mind to the entire executive management team and company owners on any given normal day.

Ouch.

Sorry.

Here's a little sugar to help the medicine go down. Not only is the disadvantage undeserved, it's unintended. Executive managers and company owners don't keep their out-of-site project managers out-of-mind on purpose. It's simply a natural human reaction that results from the daily separation and the constraints placed on everyone in this business. There's no need to take it personally or be offended by the facts. Facts are information. Information creates awareness and thus the opportunity to do something about it.

THE ROAD LESS TRAVELED

If your desired final destination is executive management or corporate leadership in a medium-to-large construction company, or you wish to start your own firm someday, consider wandering off the beaten path for the first twenty years of your career—the time that most people dedicate to becoming a great project manager and maintaining prominent status there. Bouncing around the road less traveled will give you a more well-rounded set of skills, including (but not limited to) project management, and render you much more qualified to do all of the other things required to help lead a successful construction company.

That's why many of the industry's MVPs were never full-fledged project managers, or did not stay there for very long. Many sensed early on that the time investment required to be a great project manager comes at the cost of too many other experiences required to succeed as an executive manager or corporate leader. They either figured it out on their own, or were told by another MVP, that managing a project is not the same thing as leading a company. As previously shown, it's an entirely different thing. And since they have their eye on a higher prize, they seek positions that'll expose them to the people and experiences they'll need to fast track their path up the corporate ladder and better prepare them for success when they get there.

What does the road less traveled look like?

If it doesn't run through multiple project management roles, then where does it run through? It runs through a variety of entry-level and mid-level positions, alternating back and forth from the field to the main office, exposing you to as many different activities and skill sets as possible. For example, many of our industry's MVPs spent time bouncing around in several of the following roles during their first ten years:

- Field engineer
- Assistant project engineer or project engineer
- Assistant superintendent
- Preconstruction manager
- Assistant estimator or estimator
- Assistant purchasing agent or purchasing manager
- Cost engineer or cost and fee projection manager

It's easy to see why this makes good sense. Imagine the well-rounded foundation created by spending approximately two years bouncing around five of the above things, rather than ten years on a dedicated path toward project management (or any other *one* thing). Even if you decide to pursue excellence in project management, it's important to understand that you'll achieve excellence faster with a well-rounded foundation.

As you make choices along the way, keep some of these best practices in mind. If at all possible, serve on at least one sizable project as a field engineer and another as an assistant superintendent. You'll learn more about how to actually build a building and a lot of other critical insights into how the industry works, as referenced in chapter two. Similarly, it's best to obtain both project engineering and cost engineering experience, but if you ever have to choose between them, take the cost engineering experience, without any hesitation, because it will expose you to multiple project financials and teams instead of one. Developing relationships with as many team members as possible is critical, and being a cost engineer is a great place to start. Lastly, if you get a choice between another project engineering role and a spot in estimating, take the estimating experience hands down.

The estimating department is a powerful place to learn many important things, if you're wise enough to take advantage of it. Not only how much things cost, but also how those costs are compiled and analyzed for completeness and risk—and most importantly—how they're presented and interpreted up and down the food chain. Estimating experience also makes you more empathetic to both subcontractors and project owners, from your vantage point in the middle, and affords you the opportunity to develop relationships that can have a powerful impact on your career. Plus, similar to cost engineering, you get to touch many projects simultaneously, instead of one at a time, so you interact with more people and learn more, faster.

Because they bounced around the road less traveled, many of the industry's MVPs cannot claim to be great project managers (or superintendents) but they carry sincere respect for them and understand their critical place in any construction company's corporate structure. Some were simply too curious about all of the other important aspects of running a successful construction company to retain a singular project-focus. It's a sacrifice that some highly ambitious construction professionals are not willing to make. So they strive to be well rounded, spending meaningful amounts of time in a wide variety of positions, on both sides of the buy-sell equation, knowing that it renders them distinctly valuable.

If you decide to choose it, here's what else you'll discover along the road less traveled. Even more important than the well-rounded competencies to be gained, you'll develop a strong sense of empathy for your colleagues in all departments, and a deeper understanding of the interconnectedness of everything. This grooms a more natural company-focus versus project-focus over time. It teaches you how to think at 30,000 feet above the details of any particular project, focusing you on the bigger picture, interdepartmental relationships, and company-wide initiatives designed to make things better for everyone. Above all else, if you pay close attention, you'll learn how to articulate your company's unique value proposition for any client or prospective client—to help win more work—which is arguably the single most important responsibility for any business owner, principal, or senior executive.

INCH TOWARD THE DEAL

Whether you took the beaten path and maintained a singular focus on project management, or bounced around the road less traveled to establish a more well-rounded foundation during your *first* ten years, the MVPs will tell you that it's wise to inch your way back up the project life cycle in your *second* ten years. It's critical for executive managers and corporate leaders to have experience in preconstruction, estimating, and business development. Exposure to these positions will be pivotal to your career for two reasons.

First, they take you successively closer and closer to the deal, making you directly responsible for activities involved in generating revenue, such as:

- Strategic marketing and business development planning

- Identifying and articulating your company's unique value proposition and differentiators

- Nurturing relationships with industry players and influencers

- Nurturing relationships with existing clients and prospective clients

- Qualifying project opportunities

- Answering RFPs and RFQs

- Making sales presentations

- Developing cost and schedule estimates at various design phases

- Navigating municipal approvals and expediting permits

- Negotiating contract terms

The industry's MVPs are almost always intimately involved in revenue generation in some capacity. They recognize its primal significance. There is no single activity more important to a construction company than bringing in work. Without new work, nothing else happens and no one else is needed.

Period.

End of discussion.

Second, when you work in these roles, you're stationed in the main office and you're therefore afforded the opportunity to interact heavily with your company's executive managers and corporate leaders on a daily basis. That puts you in constant contact with the big-picture decision-makers, which is crucial if you intend to join them.

Simply put—you have to be in the loop.

Bluntly put—you become who you hang around with.

THE MYTH CREATES OPPORTUNITY

Many project managers and other young professionals fail to recognize the distinction between managing a project and leading a company, or they downplay its importance out of ignorance. Others fail to acknowledge the significant differences between people who are naturally project-focused and those who are more company-focused and relationship-focused people.

Awareness of the project manager myth is therefore a huge opportunity for construction professionals who may have a wider range of interests and capabilities beyond pure construction management activities. Most people still believe in the myth and therefore compete heavily with one another to be the next big project manager on the next big project. So many of the industry's MVPs separate themselves from that crowd by pursuing a variety of other positions within our industry, many of which better groom the company-focused skill sets that are needed to reach the executive management and corporate leadership level, and succeed at them.

This formula works pretty well. It's smart to get really good at a lot of things, instead of being great at the one position where the supply often exceeds the demand. Your resulting company focus and relationships focus will position you for advancement and leadership opportunities far ahead of most of your peers.

No matter where you are right now in your career, you can pull up on the rudder and begin your ascent toward 30,000 feet. Start by implementing the following two strategies:

1. Continue working diligently in your current position and do not drop the ball in any way; however, in addition to your current responsibilities, take a 360 degree view from your desk and start identifying other company processes and work flows that you have not yet been exposed to or know very little about. Inform your boss that, although you're dedicated to excellence in your current position, you would also like to get involved in some extracurricular activities that will expose you to other department functions and company-wide objectives. This will convey that you appreciate the bigger picture, and your need to learn and earn your way to the next level. Here are some ideas that span a variety of different company activities. Try one or two of them to jump start your momentum and then see what other ideas emerge for you:

    A. Ask your company's chief estimator for permission to shadow an estimator during the next major bid and assist in minor tasks.

    B. Ask your company owners, or whoever heads up business development, for permission to attend an upcoming meeting with a prospective client—just to observe. Start wiggling your way into the sales loop, so you can learn more about what projects are coming up on your company's radar—and ask to sit in on the next practice session for an upcoming sales presentation.

C. Challenge yourself to find three appropriately experienced subcontractors that have yet to work for your company and guide them through the prequalification process.

D. Ask your controller or accounting manager for permission to shadow someone in accounts payable during a batch run preparation, to learn more about how invoices and other bills get paid (or not paid).

E. Ask whoever heads up purchasing or preconstruction for permission to sit in on an upcoming meeting with a subcontractor to finalize a (major) contract.

F. Ask the superintendent on the largest (or the most complex) project in your company for permission to attend his next subcontractor coordination meeting, and then preferably continue attending them on a regular basis so you can observe how the meeting content evolves over the life of the project.

G. Take a look at your company's last couple of RFQ and RFP submittals, to see how your company's value propositions and differentiators are being presented in the marketplace. From your perspective, provide any suggestions for improvement to the business development team.

H. Read through all of the client testimonial letters, ratings, and reviews that you can find for your company. Go way, way back to the beginning. Dig deep to find out what your company's clients had to say about their experiences working with your company. Single out a couple of them that seem particularly glowing, and sit down with your company owners (or whomever will know the most about that project) and ask them to tell you more about it. Find out some of the stories around WHY that glowing letter of appreciation was received. In summary, start learning your company's success stories, in as much detail as possible.

I. Many companies create special committees to work on process improvements and other strategic planning initiatives. Find out if there are any you can join. If there are none, take the initiative to form one on a topic you are passionate about, or one that will help your company address a problem or seize an opportunity.

2. Make a commitment to yourself that your next position will be different from the one you have now.

For example, if you're currently an assistant project engineer or project engineer, then your next position should be outside of that track. You can always come back to it later if that's what you ultimately want, but exposure to other things will only make you a better project engineer or project manager in the long run. If you're currently an assistant superintendent, then pursue a project engineer position next. If you haven't yet worked in estimating or preconstruction, see if you can go there for a while. If you're further along in your career and you believe you have what it takes to make the switch from operations to business development, see if they'll give you a shot at it. You get the idea. Expand your horizons.

As you purposefully try new things, remember that it's ideal to bounce back and forth between the field and the main office, work with as many different people as possible, and touch as many different departments and work flow processes as possible.

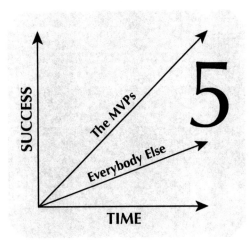

BE THE "I" IN TEAM

The idea that there is no "I" in team is one of those phrases that sound really good when you say it. It's certainly politically correct and there's something about it that feels right, maybe even a little noble, rolling off your tongue. Surely, it's something a good team leader or team player would say—isn't it? Surely, it makes your team more likely to win—doesn't it? Lots of well-wishing people certainly think so and act accordingly.

As if the concept is axiomatic.

But it isn't.

It's actually one of those little white lies that are told with good intentions. Most team leaders are comfortable telling it, because most team members are comfortable hearing it—as a promotion of unity and the stereotypical definition of successful teamwork. Unfortunately, the phrase is unintentionally misleading and it sends out the worst possible message, particularly in business and other highly competitive environments. Fortunately, the MVPs in the commercial construction industry do not buy into the message. They are not fooled. They see right through the dangerous implications of this popular phrase, to the underlying truth about successful teams, and they lead by example as highly respected "I" players. And they do it with grace and integrity.

Want what they have?

Learn how to BE the "I" in team.

GET SOMETHING STRAIGHT FIRST

Before diving deeper into what it really means to be a successful "I" player, get something straight in your mind first, so you can make the paradigm shift without concern. Here's why the idea that there's no "I" in team sends the wrong message. The phrase implies that there is no room for "I" players on a team, because "I" players are individualistic and me-centric people who are more likely to promote their own best interests above those of their team. By reverse logic, it also implies that "we" players are ideal team players, because they're less individualistic and would never do such selfish things. Again, the phrase is well intended, but the implications encourage the wrong behavior—the opposite of what's required for success.

First of all, the notion that you're either an "I" player or a "we" player is false. The MVPs are both. They are highly successful individuals who are focused on exceptional personal performance in their assigned positions, who then choose to play for teams where their skills, talents, and accomplishments are in harmonious demand—meaning the team they play for needs what they

bring in order to reach its goals. The whole thing about "I" players putting their own best interests above those of their team (or
company) is not applicable when everyone's interests are authentically aligned. When you perform the "I" part with integrity
and professionalism, and you're playing for a team that wants the
same things you want, the "we" part comes naturally by default.
So the authentic "I" players are both.

The self-absorbed ego-maniacs who are completely looking
out for themselves, with blatant disregard for the team's objectives, and couldn't care less about how their decisions affect
others are not "I" players. They are another word that starts with a
capital A. It's unfair to the legitimate "I" players to lump them in
with those guys. They get their own category entirely and nobody
likes to play with them.

Second, the implication that "we" players are ideal team players is also false, because some of them are lacking the necessary
"I" player strengths you're about to discover. When compared to
"I" players, the "we" players are less likely to be both. Also, being
a "we" player does not necessarily translate into added market
value to your team. That's why "we" player behavior is secondary.
It certainly translates into being accommodating, considerate,
supportive, and all kinds of other positive things, but it doesn't
automatically translate into added value (i.e., client satisfaction,
revenue generation, increased margins, victories, etc.).

In reality, the most ideal players on any team are simply the
players who add the most market value, period—the ones who
move the entire team closer to its objectives and increase the
value of the organization. That's where the term MVP comes
from in the first place. The most ideal team players are the *most*

valuable players by definition. They play a pivotal role in their team's success. Remove them from the team and the team's ability to perform suffers. Not because you removed their position or function, because you removed them *personally.* The value brought forth by the MVPs is intrinsic to them as individuals—as "I" players—and that's why everyone pays top dollar for them.

THE REAL "I" PLAYERS

Consider that cleared up. The "I" players have been given a bad rap by the proverbial bad apples. It's time to redefine them, so everyone can get very clear on exactly what it means to follow the industry's MVPs and be the "I" in team. All highly successful "I" players in the commercial construction industry are:

- Individuals
- Influential
- Inspirational
- And the truly elite among them are: *Iconic*

BE AN INDIVIDUAL FIRST

The MVPs are highly successful individuals, in and of themselves, independent of any team. This is the bedrock of a successful "I" player. They identify themselves as individuals *first* and members of a team *second*—not because they deem the team to be less important, don't fall into that judgmental trap—but because they understand intuitively that each team member's individual success drives the overall success of the team.

By focusing inward first, and holding themselves accountable to higher and higher standards of performance and personal achievement, they are protecting the best interests of the team. This shows tremendous respect for their teammates. In short, they pull their own weight and then some. They do more than they are expected to do (in terms of their job description) by raising their own bar higher and higher. They consistently increase their personal market value and industry impact. However, and this point is key, they're doing it in constant alignment with their company's best interests—meaning the "more stuff" they're always doing to increase their own market value and industry impact is something that increases the company's market value and industry impact at the same time.

That's why you see the real "I" players primarily focused on what appear to be external initiatives versus internal ones. It's not that they don't care about internal initiatives (i.e., streamlining an accounts payable process or chairing the annual holiday party committee)—they do care and appreciate that those things are being handled by other respected team members. They're just focused on achievements that increase their market value, which in turn, increase the company's market value (i.e., winning a coveted industry award, being appointed to a strategic board seat or presenting a keynote speech at an important industry event, etc.).

This is a powerful distinction that many "we" players miss. The "I" players maintain this behavioral prioritization as clear demonstration of a very important business principle:

*As each employee increases his or her own indus-
try credibility and value in the marketplace, the entire
team's industry credibility and value increase.*

It's all about alignment. The above principle is particularly applicable to the commercial construction industry, which is highly
competitive and sold (in large part) on the collective reputations
and credibility of each proposed team member for any given
project. Therefore, everyone is trying to attract and retain construction professionals with the most impressive personal resumes
possible—because their resumes are then used to pursue and win
work. This is one of the primary reasons why it's essential to be a
highly successful individual (first) in construction management.
The company you work for literally needs to be able to sell you
and all of your accomplishments, so the MVPs work very hard to
give them something differentiating to sell.

The key is to be a highly successful individual who happens
to participate in a team sport. Do not define yourself by the team
you play for, or allow the team to define you. Some people need
that kind of structure and support and look for the team they play
on to provide it. Highly successful individuals, who happen to
play on a team, are the ones who create the structure and support
the others need.

Make sure your personal performance on the team is stellar.
Constantly raise your own bar—be fiercely competitive with yourself. Go after every personal credential that has marketable value.
Pursue industry awards and recognition that extend credibility
and good will to your company. Become an expert individual
player. Make you the best you that you can be. Ironically, it's the

best training and preparation for team play—because it gives your company something to sell.

When in doubt, remember the flight attendant mantra: *Put your own oxygen mask on first.* It's the only way to maintain your capacity to assist others.

BE INFLUENTIAL

There are some people in the world who get things done. They know how to make things happen. The MVPs in the commercial construction industry are like that—but rarely through solo effort. More often than not, they make things happen by positively influencing and encouraging other people around them.

When the MVPs advise, people listen to them. When they perform, people emulate them. When they sell, people buy from them. Earlier in the book, you learned about some of the industry's MVPs who possess these talents. Remember Steve Sales from chapter one? His relationship with Mike Maintenance at Super Schools created strong influence over Mike's decision-making during the contractor selection process—which resulted in Steve's company winning several projects a year at Super Schools. Another example was illustrated in chapter two. Highly successful project superintendents who are magicians in the field and masters of subcontractor coordination are "I" players because, in addition to their superior construction knowledge, they're incredibly influential people. They accomplish seemingly impossible feats by moving large groups of people toward a common goal—but not through force or threat. They do it through their powers of persuasion and influence.

The root of all positive influence is trust and respect. When you are trusted and respected in the commercial construction industry, your influence begins.

BE INSPIRING

The MVPs are so good at what they do they set an inspirational example for others to follow. They maintain relentless pursuit of excellence in their personal performance, marketable achievements, industry impact, and myriad strategic initiatives designed to extend their credibility throughout the commercial construction industry at large and their local communities. They operate under the philosophy: *The more I excel and shine in my position, the more I help my team win.*

Others see this. They see the MVPs excelling and gaining recognition for their achievements, and it encourages them to follow suit, increase their own aspirations, and up their own games. The ability to inspire others to higher levels of performance is one of the inherent values of hiring "I" players—when companies invest in one, they usually end up with a few more over time.

To join the ranks of the MVPs, you must lead by inspirational example. Whatever you want your teammates to do, you must do. Showing always beats telling. Show them what it means to be a highly successful individual who happens to play a team sport called commercial construction. What does that look like? The MVPs are less directional and more inspirational. They exemplify the exceptional behavior the company needs to succeed in the industry's highly competitive environment, show everyone how it's done, and then lend guidance and encouragement to those with potential to join them.

BE ICONIC

The truly elite "I" players in any industry are iconic individuals. They are remarkably well known and admired for their unique excellence. They are literally in another league when compared to others who do what they do.

When someone is deemed to be an icon, it's more than just a really big compliment. The definition of an icon is a symbol for something. So when you hear someone say that Richard Feynman was an iconic theoretical physicist, they're really implying that he symbolizes what it means to be a theoretical physicist because he was the epitome or ideal embodiment of one. Being iconic is way above being really good. Icons set the bar by which others are measured, and they keep raising that bar.

Some of the MVPs in the commercial construction industry have reached iconic status. Everyone knows who they are within their geographic area of influence—be it national, regional, or strictly local. They're so highly respected for their unique excellence in their line of work, that they literally personify their line of work. You can identify the icons in your respective area by playing the word association game. Try it. What name do you think of when you hear the words *Hospital Construction Superintendent* or *High-Rise Structural Steel Expert?* If you've been around for a while, chances are there's one definitive name (or maybe two) that pops into your head quickly when you think of phrases like that—because those people are the iconic players in your neck of the woods.

To join this elite circle of industry icons, you need your name to pop like that, too—and today's fiercely competitive environment will force you to do much more than great work to make it happen. Prepare to add two part-time jobs to your primary full-time job, because effectiveness in two more things is quickly becoming the deciding factor between who tops the MVP list and who remains really good.

Ironically, nobody ever talks about either one of them.

PART TIME JOB NO. 1: GET A LITTLE FAMOUS

When's the last time you Googled your name? Better yet, Google your name plus the word *construction* and see what you find. More to the point, see what other people will find when they go looking for information about you. What do all the major search engines reveal about your construction career in 0.42 seconds? If the only relevant link in two million results is your sparse resume on LinkedIn or photos of your family and friends on Facebook, you're being out-played by the MVPs who are dominating the fame game with established industry credibility all over the Web.

Playing the fame game is neither facetious nor boastful or attention seeking. Obviously, this is not a reference to boys and girls having too much fun at the Jersey Shore kind of fame. It's simply an easy-to-remember phrase meant to drive home the point that it helps to be a little famous within the industry you serve, assuming of course that your recognition stems from positive contribution. Replace the phrase with personal public relations if you must. No matter what you call it, do it so you can reap the Google love that makes you and your industry street cred searchable in a fraction of a second.

> *Why is being searchable so important? Because searchable is the new recognizable. Being on TV, in the movies, or print media is no longer the only passport to public recognition. Searchable is the new famous.*

The Internet and mobile media in general have become powerfully polarizing. They work to your immediate advantage or disadvantage—there's very little neutrality left—because the ubiquitous nature of today's technology creates expectation for information. Everyone can research anything or anyone at any time, in a few simple clicks, and they do so in expectation (vs. hope) of finding the exact information they need. With potential corroborating evidence describing exactly who you are and what you've accomplished available online 24/7, the absence of it doesn't just result in a missed opportunity to create an excellent first impression, it also casts doubt that you're a prominent player at all—because if you were, the Internet would prove it.

So being searchable is no longer a good idea—it's a must. Remember when this phrase, *"It's not what you know—it's who you know,"* evolved into this one, *"It's not about who you know—it's about who knows you."* Well, thanks to modern technology, it's evolving again and now it goes something like this, *"It's not about who knows you—it's about who knows **of** you."*

Many of our industry's MVPs go to work every day within a small circle of people; therefore, the number of people and companies they interact with on a regular basis is localized and relatively low. That means the number of people who've actually met and personally know these MVPs is a small group relative to the industry as a whole. What's interesting, however, is the (much higher) number of industry professionals who can legitimately claim to know *of* these MVPs.

Even the perception of what the phrase *"I don't know him personally, but I know of him"* is changing. Before the Internet revolution, it used to mean that you've heard that person's name before and maybe a few bits of commonly known facts about that person. Now you can know of someone whom you've never met in so much more intimate detail. For example, you can connect with a construction professional on LinkedIn and then read all about his background. You can read his blog or other published content. You can watch him give a keynote speech or participate in an industry panel—either in person or in an online video. You can register to attend a webinar he's hosting or listen to a recorded podcast. You can follow him on Twitter and catch the 411 on his professional highlights—and the list goes on and on. There are many ways to be a voyeur of someone's career in the commercial construction industry.

And the point is people do.

So if you want to further your career and make more money off of all the stuff you know—meaning all of your knowledge and experience in the industry—then you have to share it. You can't hoard it. In order for people to be impressed with your knowledge and experience, they have to know you have it. And the only way to let people know that you have it is to show it off in various public ways. Then you have to prove that it happened with photos and videos, and get everything up onto the Web so people can find it in Google's famous fraction of a second.

The last piece of the "get a little famous" puzzle is to make sure that you have relevant people vouching for you with recommendations and testimonials that are visible. You want to get as many quotes, ratings, or testimonials that you can from those who've worked with you, attended your events, or read your publications and post them online where people can find them. This ensures that someone who is researching or qualifying you can:

- Find your claims of relevant commercial construction knowledge and experience

- Find proof that what you claim is true

- Find supporting evidence that your knowledge and experience is valuable to others

All of this probably sounds like a lot of work—because it is—but the days of expecting a well-written resume to get you where you want to go are long gone. The MVPs are harnessing the power of celebrity all the way to the top. You don't have to be a global phenomenon like Bill Gates or Oprah Winfrey, but you do need to get a little famous within the commercial construction circles that count.

PART TIME JOB NO. 2:
BUILD YOUR OWN BRAND

While you're getting a little famous, it's imperative to build your own personal brand. All successful teams or companies have a brand, and their best "I" players have their own personal brands. In fact, some "I" players are so influential, inspirational, and iconic their individual brand literally shapes and defines their team's brand or their organization's brand.

No icon illustrates that point better than Michael Jordan. Few would argue against the fact that Michael Jordan's personal brand dramatically shaped and defined the Chicago Bulls' brand. It still does and he's retired—that's how powerful his personal brand is within the marketplace of professional basketball and the sporting world. You could throw Derek Jeter into the same boat.

Need more convincing from outside the sporting world? Steve Jobs built a personal brand that defined Apple, Pixar, and every other company he touched. His personal brand still defines Apple after his death. His power and influence are transcendent. Bono has worldwide influence and a personal brand that rivals U2—one of the most successful bands of all time. He probably never intended that; nonetheless, his personal brand has him headed for a Nobel Peace Prize, with his loyal bandmates and millions of fans in tow. Elon Musk is another fantastic example. No matter what team he joins or creates, his mere presence defines it. Study the success of PayPal, Tesla, SpaceX, and Solar City and you'll see Musk's personal brand of worldwide, paradigm-shifting, disruptive technology genius stamped on everything.

Now, pull back from those household names, they're just meant to drive home the point. Focus in on wherever you are right now inside the commercial construction industry. More and more of the industry's MVPs are starting to build their own brands, and their brands are shaping the companies they start or the companies they join. It's time to decide who you are, what you're passionate about, and what you're willing to stand for or promote—inside the industry and out—so you can brand it in the biggest possible way.

GO ICONIC OR GO HOME

You can now move forward in confidence knowing that the "I" players on any team are neither tolerated nor problematic—they're essential. They're exceptionally talented and highly successful individuals (first) who positively influence and inspire their colleagues and other industry professionals. And the truly elite among them are the icons of their industries.

Step by step, work your way toward iconic status.

It's the key to winning.

As a general rule, there's at least one truly iconic player on virtually all successful teams of people and truly great teams have several. It is rare to see a winning team with no stand-out, iconic players. If that weren't true, patriotic fans wouldn't still weep with pride over the 1980 U.S. Olympic hockey team's shocking victory over Russia to win the gold medal. The world watched a team without any iconic players defeat a team with many, a feat so rare it's revered as one of the most spectacular events in the history of all sports. It's a huge exception to the general rule, and that's why everyone remembers it to this day.

As previously shown, the commercial construction industry not only falls into the general rule, but the highly competitive landscape actually increases the *necessity* for iconic players. Each company is reliant upon their iconic players to bring in work with their proven track records and their bona fide street cred—and without a steady stream of work, no other players are even necessary. Standout success now demands iconic stature.

In conclusion, take this last bit of wisdom with you:

> *The MVPs have learned the hard way that it takes courage to stand out as a powerful, iconic individual in a construction company, for the same reason it takes courage to do it in high school. It threatens the comfort and security of those who are not themselves powerful, iconic individuals.*

If you believe you have what it takes to join the MVPs at the top, go iconic or go home, and prepare to navigate both the admiring *and* cynical crowds that will surround you. Your potential can only be reached when undeterred by criticizing minds. As you make the paradigm shift and look around, you'll see the industry's MVPs standing courageously tall as the "I" in team.

Be an individual.

Be influential.

Be inspirational.

Be iconic . . . or be forgotten.

Before you begin this journey, get a feel for where you stand right now with a simple exercise. Pretend you are someone related to the commercial construction industry who needs to conduct research on you, or simply wants to learn more about you for whatever reason. Assume it is a legitimate industry context—perhaps they met you at an industry event, or someone spoke highly of you to them, or your resume was floated to them by a recruiter, or your name was mentioned in reference to a possible project, or your name simply came up in casual industry conversation with a colleague. Assume this person knows nothing about you, but would like to learn more. In this day and age, what's he likely to do first?

Continue to pretend you are this person in need of more information about you and your career in the commercial construction industry and Google your name. As mentioned earlier in the chapter, if your last name is a very common one, try Googling your name plus the word *construction* or the name of your company. Now, looking through this person's eyes, what do you learn about you? Does it do you justice?

The first two pages in Google are the only two that count so they need to tell a good story about your career. They need to explain three things: (1) who you are, (2) what you do, and (3) how well you do it. Most people have at least something on the first two, but fail to address the third with anywhere near the level required to differentiate themselves or their companies in the marketplace. Take a long hard look at the results right now, or lack thereof, so you can make an honest assessment of where you stand in this long and deliberate journey.

Next, plant a conceptual seed in your own mind and let it grow strong over the years to come, until it becomes habitual. From now on, whenever anything significant happens in your career, ask yourself, *"How do I get evidence of this onto the Web?"*

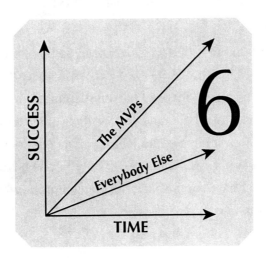

PEOPLE ARE NOT
SHOP DRAWINGS

When you treat people like a set of shop drawings you destroy your relationship with them, along with any chance you may have of being an effective leader in the commercial construction industry. The same holds true for new ideas; you can't treat them like a set of shop drawings either. Fail to embrace this strategy and you'll never reach the executive ranks or make a complete mess when you get there. Know this for sure:

> *There's a time and place to look for everything that is wrong or missing, and there's a time and place to look for everything that is right and possible.*

The MVPs in our industry learn how to do both, but they excel in the latter because the former is a lousy way to lead. To become the requisite company-focused person identified in chapter four, you must pack your bags and follow the yellow brick road to the magical land of people and ideas. Great construction leaders take up permanent residence there. They become fruitful farmers and protectorates of this land, fostering a place where people and their ideas are free to flourish under watchful eyes that are focused on what is right and possible.

How do they do it?

The transition from project-focused skills and behavior to company-focused skills and behavior requires many things, for which several books could be written; however, keen awareness of the following insight just might be the most critical requirement for the personal success of any leader in the construction industry and the well-being of the company he works for:

> *You must become aware of your Pistol Pete and why he is no longer your trusted friend.*

MEET YOUR PISTOL PETE

All construction professionals have a Pistol Pete living inside their heads. His job is to spot errors and omissions and potential problems. He likes to be the first one to point out why someone is wrong or why an idea won't work, because he's relatively slow to trust new people and new ideas. He loves to argue and he runs excellent defense. Picture a big linebacker right before he makes a tackle—he knows his job is to prevent anything from getting past him.

Your Pistol Pete is loud and proud, and as you gain more experience in the industry, he gets louder and prouder. Oh, and he's trigger-happy, too—which means he's quick to object, judge, and criticize. Basically he lives to shoot holes in everything and that's how he gets his name.

EVERYBODY HAS ONE

Construction professionals are literally trained to develop a sharp-shooting Pistol Pete. This behavior is encouraged and pervasively ingrained into the minds of construction managers everywhere. Right out of the gate, almost all entry-level and mid-level positions require this mentality and the resulting behavior in order to be successful. The industry therefore molds and demands people to remain on high alert for everything that may be missing or wrong or potentially problematic with nearly everything they see, hear, or touch every day.

Every construction professional has a Pistol Pete working hard on all kinds of things that are familiar to you. Take shop drawings as an example. One of the most common tasks in construction management is the review of subcontractor shop drawings—and what does a good project engineer, project manager, or superintendent do with these shop drawings? He calls his Pistol Pete to attention and goes to work scrutinizing every page, meticulously looking for anything that is wrong, missing, or potentially problematic.

When your Pistol Pete does a good job reviewing shop drawings, you typically feel pretty good about yourself and often get
acknowledged for it, which reinforces your behavior. Now let's
extrapolate the point further. Pretty quickly, your Pistol Pete gets
better and better at his detective skills, and therefore you become
more and more adept at pointing out everything that's wrong or
missing or potentially problematic wherever you go—feeling like
a million bucks every time you catch something and even better
when you catch something that everyone else missed. And all of
this serves you well, because your Pistol Pete's skills and behaviors are needed when you review construction plans, schedules,
specifications, cost estimates, contracts, RFIs, submittal packages,
change orders, and all of the other stuff that must be managed
throughout the construction process. Virtually everything in the
project-focused world strengthens your Pistol Pete and that's why
everyone who survives past the first couple of years in this industry has one.

Your Pistol Pete is indeed productive and valuable.

Up until a certain point . . . then it's no longer wise to listen
to him.

PISTOL PETE CAN GET YOU INTO TROUBLE

Here's the thing. As your Pistol Pete strengthens, somehow,
somewhere along the line, he becomes something more than the
part of your brain that you tap into when you review shop drawings and the like. If you're not careful, he becomes an everyday
attitude—a way of approaching your work with suspicion, doubt,
and an authentic sense of responsibility to point out perceived

shortcomings, spot potential trouble, and preemptively shoot it down.

Right or wrong—ask questions later.

In the case of our industry's MVPs, this divisive attitude lies subtly in their peripheral vision, where it should be, under control and situational discretion—but for most, it tends to direct their entire line of sight, regardless of the activity at hand and whether or not the attitude is appropriate. It may be unintentional, but it's largely why many construction professionals treat people and ideas like a set of shop drawings, and it gets them into trouble.

Nowhere does this cause more trouble than in the executive level ranks, where construction professionals *must* migrate from managing construction to leading people, relationships, ideas, and change. As previously discussed in chapter four, leading a company is very different than managing a project. It requires a well-rounded set of skills and behaviors to succeed and your Pistol Pete doesn't have all of them. In fact, he has much of the opposite. At this point, instead of being productive and valuable, he becomes destructive and costly. Let's break it down and see some of the reasons why.

PISTOL PETE DISCOURAGES PEOPLE

Reread the opening sentence of this chapter: *When you treat people like a set of shop drawings you destroy your relationship with them.* It's true in virtually all relationships across the board, whether personal or professional, because the behavior discourages people and drives them away from you. In short, nobody likes to be around perpetual fault-finders. When people act like that, they're off-putting and uninspiring and exhausting. Leaders need to breathe life and creativity into an organization, not suck all the oxygen and self-expression out with constant scrutiny.

Think back to the shop drawing analogy for a minute so you can see the exact behavior at play. When you look at a set of shop drawings, you literally mark up everything that is wrong, missing, questionable, or potentially problematic on the drawings in front of you. Notice that you do not indicate what is right on the drawings, because that's not your job in reviewing them. What's right is not acknowledged at all. The task at hand gives you permission to ignore what's right when communicating the results of your review to others. You are simply a trouble-shooter on the lookout for trouble and a fault-finder on the lookout for faults.

When you take this heavily engrained behavior and apply it to everyone around you, especially those who work for you, it gets you into trouble. When you're constantly focused on what's wrong or missing or potentially problematic in a person or team of people, rather than on what's right or possible, they may take it for a while or even temporarily benefit from your criticism if they perceive it to be constructive and well-intended. However, at some point, if you fail to acknowledge what's good and potentially great, they may feel under-appreciated and disrespected. They may even feel like they're failing. Even if you actually do respect them and deem them to be valuable, they may not realize it based upon the words or feelings you are actually communicating. To use a construction metaphor, it makes people feel like a human punch-list when you're constantly taking their inventory in terms of what's still not done or still not acceptable to your standards. You get the idea.

In fact, reading this has probably reminded you of a time when you were treated like that by a co-worker, boss, or authority figure, so you likely know how it feels firsthand. Maybe your current boss treats you like that right now. Odds are pretty high that he does, at least some of the time, because the Pistol Pete behavior is so prevalent. Conversely, you've probably treated others like they were a set of shop drawings or a punch-list, too, albeit unintentionally. Some of this behavior may come from personal experiences unrelated to your career in the commercial construction industry. It might be part of the baggage you brought with you when you got here; however, be enlightened further by connecting the dots to see where the tasks and activities in your career reinforce this behavior and strengthen the bad habits.

Now that you know (at least) one of the reasons why some construction professionals rub everyone the wrong way, you can be cognizant of it and guard against letting your Pistol Pete demotivate everyone 24/7.

PISTOL PETE PREVENTS CHANGE

Try exploring a stupid question. Have you ever noticed how painfully slowly this industry adopts new ideas or change? From new construction materials, to new installation methods, to new ways of doing business, this industry is pretty slow. And technology! What's that? In some circles, construction professionals are downright infamous for their snail's pace adoption of virtually all mainstream business technology, including all communication, productivity, and sales tools. From fax machines, to personal computers, to software, to cell phones, to the Internet, to email, to mobile apps, to you-name-it, construction is behind the adoption curve relative to most industries and some experts would say it's more like dead last.

That leads to a less stupid question. Have you ever stopped to think about *why* the industry is slow to adopt new ideas and promote change? There's probably more than one answer at play here, but your first thought might be the notion that *"It's hard to teach an old dog new tricks"* and therefore you place some (or most) of the blame on the fact that many decision-makers in the industry are old dogs who grew up before the age of personal computers and the Internet. Or maybe they're simply much handier with a hammer than a keyboard.

Yes—that could all be part of it.

But could it be that some dogs, young or old, simply have an out-of-control Pistol Pete that likes to shoot holes in all the new tricks? And, worse yet, think about what happens when that dog is the alpha dog in the pack who sets the example and makes the rules for all the other dogs to follow.

Been there?

Seen that?

And what would a dog that lets his Pistol Pete shoot holes in a new trick sound like in the construction industry? I bet you can guess. You've likely heard it before, or maybe even recognize it in yourself, sounding something like this:

> *The answer is NO. We're not doing it. Just drop it.*
> *This new trick is not going to work for us. Maybe it*
> *works for other animals but we don't need it. Dogs*
> *are different. It won't be worth the time and money*
> *we'd have to spend figuring out how to actually use*

it. I know some of the big dogs are doing it, but we don't have to do everything they do, and our old tricks work just fine. In fact, I've never even seen another dog around here try it, except Sparky down the street, and he spent six months trying to make it work and got run over by a truck. All of this talk about new tricks is just a distraction from what we're supposed to be doing. Look, we're dogs. What we do is pretty simple. We sit, stay, lie down, roll over, speak, and fetch the newspaper for our Owners. I really don't think they expect us to post the details online, text pictures, or tweet them an hourly report. Arf.

If you struggled between sneers and tears reading that, you're not alone. It's hard to know whether to laugh at the preposterousness of the situation—or cry—because everyone's known an alpha dog high up in a construction company who adamantly refused to try something different. Whether it was adopting a new technological tool, creating a new position in the organizational chart, exploring a new niche or a new way of structuring the company, or anything else you want to substitute in there. Some people have such a domineering Pistol Pete that it retards positive change within their department or company, and some work unconsciously hard at preventing change entirely.

That's why you can't treat new ideas like a set of shop drawings. It squashes them before they're fully explored for their true potential. It literally kills creative thinking and the process of objective analysis. Clearly, not all new ideas will be good ones; however, if your Pistol Pete is running loose, you'll miss the opportunity to fairly evaluate them. The MVPs know that a list of potential reasons NOT to try something new doesn't automatically mean it shouldn't be tried—no matter how long the list. In the world of pros versus cons, sometimes the pros win even when they're outnumbered. The potential good of one single possible outcome can be so great that it outweighs all the potential bad things that could happen—but good luck selling that premise to your Pistol Pete. He'll argue to the death against it, and if he loses the fight and the idea moves forward against his recommendation, he may even sabotage its success just to prove he was right.

In case your Pistol Pete is arguing right now, defending how important it is to consider the potential dangers and pitfalls of any new idea—well, that's a big fat duh. Making a list of concerns is obviously warranted and the list should not be ignored; however, the concerns may just need to be proactively managed in order to minimize their risks. So, by all means, let your Pistol Pete create a list of the cons as required, and then use that list to develop appropriate risk mitigation objectives—just don't miss the point. You have to ignore your Pistol Pete entirely when it comes to making a fair list of the pros; he's not invited to that party. And definitely do not let him make the final decision. Your Pistol Pete is not capable. He'll just sound like one of those old dogs barking the sound of *resistance, resistance, resistance*—all for the sake of pointing out the one thing that could go wrong ten miles down the road, rather than acknowledging the ten things that are already right, and the hundred more things that just might be possible.

Lastly, if there's one thing the commercial construction industry needs it is creative thinkers, and truly creative people do not thrive under a Pistol Pete. Eventually, they leave. Most construction companies have enough expert fault-finders. What they need are more people who can see possibilities. If everyone wants the industry to improve and evolve at a faster pace, then everyone must prevent the collective Pistol Pete from driving the creative thinkers out, shaking their heads in frustration saying, *"Those people cannot get out of their own way."*

DON'T LET HIM DRIVE

As you make your way up the corporate construction ladder, do not let your Pistol Pete drive. He can ride along, but he can no longer drive. This awareness is instinctual for the MVPs in our industry—they get it in their gut—but for most, this awareness must be learned and adjustments must be made. For those who need help visualizing the instinct, it should sound something like this in your head:

I know I have a Pistol Pete and he's been my trusted friend. But he cannot lead here. This place is different. If I let him run this department, or company, nothing will grow or reach its full potential. My people will become disheartened and uninspired. Relationships will stagnate. Everyone's ideas will be stifled, including my own. Nothing will improve. I better retire my Pistol Pete. No, that's not the answer. I will still need him for some tasks. Maybe I'll just keep him on the side, like a consultant. Yep, that's it. Now I can call on him when I need to ask him about something very specific, but not be clouded by his tendency to find fault in everything. OK, good. Aha! Now I can see more clearly. Now I can see what's right and what's possible here.

No laughing allowed. That may have been a little corny to get your attention, but the sentiment is accurate and it represents the fastest path to the top. Even more important, it's a better strategy for accelerated success when you get there.

LEAD BY EXAMPLE

Make no doubt about it. Your Pistol Pete is real and he'll be very useful in many construction management and risk management activities. So by all means, let him out and use him when applicable. Just be mindful, as you climb the corporate ladder higher and higher, away from project-focused construction activities and into the leadership of people, the forging of relationships, the exploration of ideas, and the advocacy for positive change, you'll need him less and less. Eventually, you'll discover that great leaders rarely consult their Pistol Petes at all. When fault needs to be found, it's often wiser to delegate that responsibility to others, so you can remain positive and forward thinking.

Get this one wrong, and you'll lose to the guy who regularly tempers his Pistol Pete, so he can attract and retain creative thinkers and problem-solvers, and fairly evaluate ideas and opportunities for growth and improvement. He might be sitting right next to you, or working for another company, but either way, he'll fly right by you. Get this one right and you'll not only beat your competition, you'll become part of the solution instead of part of the problem. In so doing, you'll inspire the next generation of construction professionals to follow you, one by one, until the entire industry starts changing for the better . . . faster.

*A **final note:** If you're rolling your eyes, finding fault, looking for exceptions or otherwise shooting holes in this theory—thank you, Pistol Pete. You're demonstrating the point and failing to grasp it at the same time. Read this one again.*

There are two simple things you can start doing now to put this important strategy into action. They'll help you and others in your company to explore new ideas with more objectivity and clarity, embrace change, and ultimately become better listeners and leaders.

1. As you may have already guessed, it's time to name this voice inside your head. Just do it. You don't have to tell anyone. If you think the name *Pistol Pete* fits, then just use that. If not, try something else like *Negative Nancy, Fault-Finder,* or *Debbie Downer.* Use whatever comes to mind when you think of yourself reviewing a huge set of shop drawings, a complicated bid package, a change order, or a big CPM schedule.

Better yet, picture the way you feel when someone is trying to sell you something. There's a good chance that your Pistol Pete is triggered quickly and easily by salespeople, especially if they're trying to sell you something that you don't fully understand.

The good news is that your Pistol Pete is not actually *you* per se. It's just that voice you hear inside your head telling you to run defense, defense, and more defense. This distinction becomes clearer if you give that voice a name. It helps you detach yourself from him, so you can begin making more conscious decisions about when and where to proactively use his skills, rather than letting him react to everything all willy-nilly. Once you name him, you'll then know whom you need to silence whenever you find yourself looking for everything that's wrong, missing, or potentially problematic when you should be looking for everything that's right and possible.

2. Share this idea with your colleagues, including your boss, who might be a trigger-happy Pistol Pete and not even realize it. If you have a hard time explaining the concept to others, just tell them to read this book. Encourage everyone you work with to raise their awareness level about this pervasive, natural tendency, noting where it is productive and valuable, and where it is destructive and costly.

When you're having an important conversation or sitting in a meeting and someone's Pistol Pete gets in the way of objectivity or progress, kindly interrupt with a little smile and say something like, *"Hey, who let Pistol Pete in here?"* It may help stop the negative flow of energy and change the trajectory of the discussion. It's pretty easy to get away with that tactic, and it works pretty well, as long as the permission flows fairly in both directions. You have to foster an environment where everyone is allowed to call it out when they see it, including back on you, so that everybody's Pistol Pete can be effectively sidelined whenever he shows up uninvited with bullets a-blazing.

Start interrupting the pattern. Be the person in your department or company with a consistent open mind to learn and explore new things, with a willingness to seek the best in people and catch them doing something right. When you do this, you may initially find yourself celebrating at a party for one, because this behavior is not common in the construction industry. So what—that works to your advantage. Stand out like a refreshing sore thumb. You'll be rewarded for it someday, somewhere.

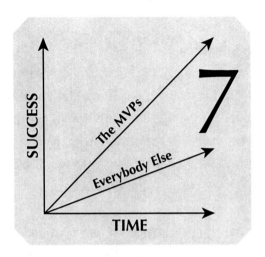

RIGHT, RIGHT, RIGHT, LEFT, RIGHT

If you want to march steadily along in your construction career, keep chanting the familiar *"left, left, left, right, left"* in the back of your mind like almost everyone else does, and you'll do just fine. It works as a good metaphor. Swap out your left foot for your left brain, tow your right brain along for the ride, and you'll fall in step with the steady cadence led by our industry's left-brained majority. After all, predominantly left-brained people pursue careers in construction management and enjoy success working in many positions that require heavy-duty left-brain power, such as scrutiny, logic, list making, scheduling, linear processing, and the organization and analysis of highly detailed information. While performing all of these left-brain tasks, the right side of our collective brains can practically take a nap.

It barely gets exercised at all.

The left brain is clearly the industry's comfort zone and re-cruiting zone. So if that's how you're naturally wired, you'll find yourself in good company. Assuming you're good at your job, your contributions will be rewarded with gainful employment and you'll be accepted into the ranks. After all, many left-brain skills are required to manage the construction process. In short, you'll fit right in.

But here's the thing. If you're aiming high and aspire to join the MVPs who really run the show, at some point, you'll have to learn how to switch gears quickly and appropriately, so you can tap more frequently into the power of your right brain. It's argu-ably much more valuable for your career. If you're atypical for the industry and naturally right-brain dominant or double dominant—which means you're very strong in both hemispheres—then switching back and forth will come easily and you'll enjoy find-ing the unexpected company of others who think more like you do. If not, you'd better take your right brain to the gym and muscle up because the MVPs are chanting *"right, right, right, left, right"* all the way to the top.

THE RIGHT STUFF

It's important to note that complete lateralization of the brain is more pop psychology than factual neuroscience. Both hemi-spheres actually work together through the corpus callosum to perform many tasks, and everyone without brain defect or injury uses both sides of their brain to function at all times. To imply that people use or think with only one side of their brain is misleading and unproductive. It should also be noted that neither side of the brain is *better* than the other. However, loosely categorizing people as left-brain or right-brain dominant can be useful when clarifying how people learn, how they think, what they tend to think about, how they solve problems, and what activities they're more likely to enjoy and excel in.

So then what's all the fuss and importance about using your right brain in the commercial construction industry? And why do the MVPs tend to be right-brain dominant or double dominant, when so many of the tasks required in contracting and construction management are governed by the left brain? The answers are fairly straightforward.

Here's what the MVPs seem to have figured out. As you go up the ladder, you'll still need a sharp left brain to organize and analyze a great deal of technical and financial information, read and write with precision, evaluate and implement processes and procedures, and other related activities; however, your right brain will have the right stuff for nearly everything else required to succeed at the top. Therefore, highly competent and ambitious construction professionals who are right-brain dominant or double dominant have a distinct advantage over the left-brained majority, whether they were born that way or worked diligently to become that way. The right stuff that's powered by your right brain includes the ability to:

- Be highly and demonstratively passionate
- Be optimistic, particularly in the face of adversity
- Empathize with others
- Recognize emotional cues

- Value and nurture personal relationships
- Think outside the box and foster creativity
- Respect and encourage ideation
- Embrace change and advocate for change
- Be highly intuitive and instinctive
- Tell engaging stories
- Act and react quickly to seize opportunities
- Synthesize information
- Recognize patterns and relationships between seemingly unrelated things
- See the whole of things vs. the parts

3X THE MILEAGE

There are many ways to effectively leverage your right brain to excel in the seemingly left-brain world of construction management. Everything from creative problem solving in the field, to killer marketing ideas, to relationship skills that turn clients into raving fans, to strategic planning, to building a strong and beloved corporate culture where everyone is authentically excited to come to work every day are all talents born from the fire burning in your right brain. An entire book could be filled with inspiring applications; however, there's one helluva good example that's worthy of a deep dive because many of the industry's MVPs are out there executing it beautifully—all the way to the bank. It's such a subtle and instinctive talent that some may not even real-ize they're doing it when they're doing it, but most work hard at it consciously. They know it's an incredibly important part of their success equation, because it literally generates three times the career mileage when you become rock star good at it:

1. Practicing it will turn you into a right-brained ninja and amp up your qualifications for executive management.

2. Applying it while serving in any project-focused role (i.e., project manager, project superintendent, and project exec-utive) will help you defy the odds outlined in chapter four and shatter the project manager myth.

3. Mastering it will result in one of the best possible bonus qualifications to differentiate yourself and your company in the marketplace, as discussed in chapter one.

It's really very simple. The biggest super power in your right brain is all about understanding that feelings trump facts, and then acting accordingly.

Really—in the commercial construction industry?

Yes—especially in the commercial construction industry.

FEELINGS TRUMP FACTS

You'll find a pot of gold in the most unlikely place. It's hiding in plain sight at the intersection of construction management and a famous quote by Dr. Maya Angelou:

"I've learned that people will forget what you said, people will forget what you did, but people will never forget how you made them feel."

Those aren't just words to live by. They're words to work by, in any industry. In the commercial construction industry, however, the quote is more than just a moral compass. It's also a strategy the MVPs use to help turn a unique industry problem into an opportunity to shine.

The problem is this. Construction managers live in a world of estimations and educated projections, coupled with the responsibility of managing unpredictable groups of people and the coordination of a an exhausting number of moving pieces and parts—that are all subject to change at any time. There are also plenty of uncontrollable factors to contend with along the way, such as project owner decisions, equipment lead times, sudden market shifts in labor and materials, and the weather.

At best, everyone does their best to make informed decisions, draft a solid game plan, adjust it as required, solve ever-emerging problems, and move the ball every day one step closer to the contracted deliverables. However, truth be told, when a construction company begins the process of building a building, the only thing you can say with any certainty is that nothing will go down *exactly* like it was planned.

They're just too many variables and unknowns. It's an imperfect process, even for the very best construction companies who employ exceptionally trained and skilled people. Therefore, despite herculean effort, construction professionals often find themselves working from assumptions, estimates, and schedules that require constant adjustment—the consequences of which then need to be effectively communicated and managed, to the highest possible understanding and satisfaction for all project stakeholders.

What's interesting is that every once in a while something that gets said or done along the way ends up being 100 percent dead wrong and way off track, but most of the time, it's all the little things that end up being "not quite exactly right" that create the constant need to manage everyone's emotions and reactions. The wisest construction professionals will tell you:

> *Success in construction management over the long haul is less about managing the consequences of being occasionally dead wrong about something that was said or done along the way, and more about managing the consequences of being not quite exactly right.*

The MVPs fight this ongoing battle with calm diligence and relentless grace. Not a battle against the project team, a battle against the imperfect process. The domino effect of all changes in scope or schedule, unknown existing conditions, economic forces and everything else, is loosely analogous to what is known in the scientific community as *chaos theory*—a mathematical framework developed by Edward Lorenz that describes how even the smallest of changes in a system can produce highly complex and unpredictable results.

Sound familiar?

Something akin to Lorenz's "butterfly effect"—the flap of a butterfly's wing in Brazil potentially causing a tornado in Texas—is almost palpable on complex construction projects. It's intensely challenging to predict the consequences of every chain reaction and effectively manage the entire project team through those consequences, particularly when working with inexperienced project owners who are unfamiliar with the nature of the construction industry and therefore legitimately nervous and untrusting of the process.

Accepting this reality, the MVPs understand that much of what gets said and done throughout the life cycle of any project will be viewed as "not quite exactly right" or fall out of line with someone's expectations somewhere. They know it's commonplace and very difficult to avoid. It's the nature of the building beast. So they hang in there, diligently and gracefully guiding everyone through the storm.

With respect to this unique industry challenge, you might say that Dr. Angelou's quote delivers a welcomed relief that most people will forget what gets said and done, since much of it will turn out to be inescapably off the original mark, even if just a little. More importantly, however, her words reveal a higher calling to remain cognizant and respectful of how you make people *feel* along the way, which leads to the real solution and its multiple rewards.

THE MVPs ARE EXPERIENCE MANAGERS

Once you understand that feelings trump facts, you can follow the MVPs who are chanting *"right, right, right, left, right"* and take a short leap to the realization that successful construction management is now all about experience management. It has become essential to manage everyone's feelings toward an ever-positive emotional state—throughout the difficult, expensive and highly stressful experience of building a complex commercial project—if you want to turn clients into repeat clients. This is no small task, and it all happens inside your right brain.

Everyone has an experience manager that lives inside the right brain, but everyone's experience manager is not equally engaged or capable of producing consistently positive results. People who are right-brain dominant or double dominant tend to be better experience managers by nature, meaning they are (1) highly cognizant of other people's feelings and the collective experience they are creating around them, and (2) they are more likely to respond empathetically and take actions to ensure the highest possible harmonious state. Conversely, others who are left-brain dominant are stereotypically less skilled at these things and therefore may be less likely (or slower) to accurately intuit everyone's collective experience. They may also lack confidence in their ability to intervene or positively influence another person's feelings, so it's understandably easier and more natural for them to re-focus back into their comfort zone of left-brained activities. Point being, some people have to work much harder than others to get really good at it.

Whether by nature or diligent hard work, the vast majority of the industry's MVPs become excellent experience managers. No matter what happens out there, somehow, someway, they're leaving their clients and colleagues with a good taste in their mouth and a good feeling in their gut. They're literally taking this unique industry problem and turning it into an opportunity to outperform everyone else. They accomplish this by being mindful of Dr. Angelou's famous quote and using their right brains to fire up behaviors that result in people *feeling* appreciated, respected, and fairly treated—regardless of all the details, facts, figures, and events of the construction management process. Here are four examples:

1. They are instinctive, keenly intuitive, and empathetic, which are three of the most important things needed to assess situations, spot discontent or confusion, maintain awareness for other people's feelings, and see things from other people's point of view—especially project owners and subcontractors. This inspires feelings of trust and good will.

2. They keep a watchful eye on the largest possible picture. They see things from 30,000 feet, sensing the health and welfare of the project as a whole, avoiding potential distractions from hyper-focus on any constituent parts. This inspires feelings of faith in leadership and direction.

3. They exude knowledgeable passion and they are demonstratively optimistic, particularly in the face of challenges and difficult odds. This inspires feelings of confidence and energizes the troops.

4. They are highly creative thinkers and problem-solvers. They see opportunities and think of ideas that most people miss or dismiss. This inspires feelings of relief and calmness when problems arise, knowing that the team can overcome (nearly) all obstacles.

At the end of the day, and the end of most projects, the way your client feels about you and your company, and the overall experience they had working with you while their building was being built, will trump the facts of whatever actually happened out there. The details will fade, but the feelings will last.

Consider yourself informed.

The highly technical, data-driven, science-like nature of the industry is largely an illusion. It's nowhere near the full story, and it's not the part that counts the most—in the end—when the final scores are tallied. With their keen intuition, the MVPs have figured out that if feelings trump facts, then experience management must trump construction management. This key insight flavors the main ingredient in their secret sauce and people eat it up and come back for more.

IF YOU'VE GOT IT—FLAUNT IT

So now you know. The best commercial construction companies in the world are looking for professionals who know the difference between managing the project and managing the client's experience. They do both very well, but they focus heavily on the latter because the former is no longer a differentiator when comparing true apples-to-apples competitors.

Think back to chapter one. Excellence in the core principles of construction management—to complete projects on time, on budget, with good quality construction—are now the *basic* qualifications required to stay in the game and make any money. For all intents and purposes, legitimate competitors are equally good at it, so it's perceived to be an available commodity. Therefore, the construction management process *itself* cannot and will not sell you or your company. No matter how good you are at it. You simply cannot get "excellent enough" at all the facts and figures and events and activities for it to mean enough anymore.

Excellence in managing the client's experience throughout the construction management process, however, is a *bonus* qualification. Not all legitimate competitors are equally good at it; therefore, if you're an exceptionally talented experience manager, you can rest assured that it's one of the best possible answers to the *"What else you got?"* question.

How you make people feel, when all of the commoditized construction management activities are unfolding, is a powerful differentiator. It straight-up matters more than anything else. It gets remembered. It gets around.

> *When all is said and done, the fact that a building gets built is shockingly beside the larger point. How everyone feels about their overall experience working with you and your team defines everyone's personal reputations and (in turn) your company's ability to compete.*

The implication here is that you'll either enjoy or suffer the consequences of your experience management skills. More than any other singular differentiator, the MVPs are taking this one to heart, following Dr. Angelou's wisdom and turning one of the industry's biggest problems into a competitive advantage. It's why they're miles ahead of their colleagues and why the companies they work for are miles ahead of their competitors.

The very best construction managers are experience managers first and construction managers second. Maintaining this priority will skyrocket your career. Working hard to become an excellent experience manager shatters the project manager myth and grooms more of the skills required to succeed as an executive manager and corporate leader. It may very well be today's ultimate bonus qualification.

If you've got it—start telling stories to prove it.

And ride it while you still can.

WHEN IN DOUBT—TRUST YOUR RIGHT

As in most industries, the MVPs in the commercial construction industry lead with their right brain almost entirely. They have an innate sense about people and things and where everyone and everything is headed, or more importantly, could be headed. Less distracted by the parts, which they effectively delegate to other experts, they see the whole more clearly. And the truly great ones—the super heroes—are idealistic and impatient visionaries. Those three right-brain characteristics bear repeating: *idealistic, impatient,* and *visionary.* The super heroes trust their idealistic nature, as opposed to allowing their Pistol Pete to shoot holes in it. They trust in knowing that time is precious and their impatience is wise. And they trust that their vision can be realized, so they bring the will to create the way.

If you're having a hard time associating those three words with anyone in your current world, that's all right. Just believe it is possible and they're out there. You never know, the magic may already be growing inside of you.

As you make your way, be inspired by the double dominant super heroes of other highly technical industries that were mentioned in chapter six—like Steve Jobs and Elon Musk. There are geniuses and then there are magnetically powerful leaders, but rarely are they found in the same person. That's what makes them super heroes. Perhaps the only thing more amazing than the capacity of those two guys' left brains is the uber-talented gifts of their right. Imagine telling either one of them that their ideas will never work in the real world, or cannot be done that fast, or that no one will ever buy them. Many tried and were proven wrong.

Start believing.

Then start chanting *"right, right, right, left, right"* wherever you go. Lead with your right onto every jobsite, into every department, up every corporate ladder, and into every boardroom. It will safely guide your left, and everyone else's.

When in doubt, your right brain is righter.

In many respects, it takes someone with a pretty dominant right brain to fully comprehend the wisdom of Dr. Angelou's words and their critical application to the business of construction management. And it surely takes some serious right brain chops—like passion, sensitivity, empathy, intuition, and the ability to see the whole above the parts—to become an excellent experience manager. So if this entire chapter reads a little *fluffy* to you, or you doubt the wisdom that feelings trump facts, you're probably in the left-brain dominant majority.

No worries.

That doesn't mean you cannot strengthen your right brain talents. It just means that it may not come as easily or naturally, or make as much sense to you at first. Try replacing the word *fluffy* in your mind with the word *strategic* to help you take everything more seriously. Remember you have no choice if you want to compete with the MVPs, as relying solely on being a really good construction manager is no longer enough.

To get started, here are three ways to beef up the muscles in your right brain, raise your awareness level, and start fast-tracking your career.

1. Read Daniel Pink's book: *A Whole New Mind*. If you find your Pistol Pete doubting the claims and theories asserted in his book, or you're not convinced they apply to this industry:

   A. Slap yourself across the face.

   B. Wake up to reality.

   C. Read it again.

   Hello, the book applies to all industries. That was his entire point. In fact, the commercial construction industry needs the paradigm shift more than many others. So it's time to become a part of the solution, not the problem.

2. Try a little exercise in experience management. Some of you will hate this. If so, let your hateful reaction serve as a sign that you may need this exercise more than others. Pick someone you work with on a regular basis, someone with whom you get along fairly well and share a good working rapport. This person can be internal to your company or someone outside whom you interact with on a regular basis. Write out a simple but thoughtful summary of how you think that person would respond if you were to ask him, *"What's it like to work with me?"* The key here is to objectively walk in his shoes, see through his eyes, and attempt to describe his feelings about working with you. At this point, you are practicing empathy. Now it's time to test it and see how well you did.

    Sit down with this person and tell him you need his help for a few minutes. Explain the exercise and without disclosing your predictions, ask him, *"What's it like to work with me?"* Give him a chance to think about it and encourage his honesty, so you can see how far on or off you were in your analysis, and what you can do to improve that person's experience with you. Not only will this be a great exercise for your right brain, you'll likely strengthen the relationship further in the process.

After completing this exercise, do the same thing with another person who is more of a question mark for you—meaning, someone whose perception of you you're not really sure of. This one will be much more difficult, but the potential rewards are greater and it will certainly stretch your right brain in the right direction.

3. This last one is a doozy. It's the most important advice an ambitious construction professional could ever receive. It can literally make or break your career. Drum roll please. Start mastering the art of storytelling. Being an engaging storyteller is the absolute best way to communicate anything that you want to be remembered, influence the way other people feel, and sell anything—especially yourself.

In closing, know that you and your career are not a bullet point list of facts and figures and summarized accomplishments. You are the thousands of colorful and compelling stories that describe the impact you've made on the world. If there's one single thing all of the MVPs in the commercial construction industry have in common—it is this:

> *They speak eloquently and powerfully in the first person. They all know how to tell and sell their own unique stories.*

AFTERWORD
A call to action for our industry's MVPs

Since graduating from Michigan State University with my construction management degree in 1990, I've been making a long-term observation. A growing number of global industries have been completely reinvented by some of the world's most daring and successful leaders—who stepped into existing business models, inspired radical change, massive growth and profitability, and began moving those industries closer and closer to their most ideal states.

Ted Turner chose the cable television industry. Jeff Bezos chose retailing. Steve Jobs chose personal computers and mobile devices. Elon Musk chose e-commerce, electric cars, solar power *and* space exploration. And the list goes on. All of our lives would be qualitatively different right now, if those people had focused

on smaller, niche industries or settled for anything less than absolute revolution.

As I continue marveling at all of their accomplishments, I realized that perhaps my paramount intention for this book is really
a call to action—to inspire and challenge the MVPs of our industry to keep pushing the existing boundaries and kick things
up another notch ASAP. The faster we up our own game, the better chance we have of competing with other global industries in
the cosmic battle to attract elite entrepreneurs with proven track
records, and our fair share of the uber-creative, uber-intelligent
young people entering the work force.

I often wonder—who will be our Elon Musk?

Maybe it's you.

As an industry, we need and deserve that level of change
agency. After all, what we do for a living is undeniably important
to the world—it's up there with water, food and energy. Nearly
everywhere you go outside of your own home was built by the
commercial construction industry. We literally fail to touch no
one.

> There are only a handful of industries on the planet
> that are truly essential to human civilization, and ours
> is one of them.

Some of the places and spaces we build are awe-inspiring.
They're tangible, beautiful works of art that stand as a testament
to human ingenuity, craftsmanship, and endurance . . . and yet,
they fail to attract the growing number of brilliant minds required

to propel us ever-forward, in pace with our global industry competitors.

Perhaps our future lies in how quickly we can make the construction process, the experience, the industry *itself* . . . equally sexy to the buildings we build.

ABOUT THE AUTHOR

Coty Fournier is a U.S. commercial construction executive and entrepreneur with 25 years of sustained success in construction operations and business development, on both sides of the owner-contractor equation. She is also the co-founder and former CEO of Jobsite123.com—the industry's first search and qualify engine—acquired by The Blue Book Building & Construction Network in May 2013. In addition to her senior leadership role as vice president of The Blue Book, she continues to serve the commercial construction industry as one of its most prominent thought leaders on executive talent development, business development strategies, effective marketing platforms and workflow technologies.

She began her construction management career as a field engineer with Turner Construction Company in 1991 where she excelled in several assistant superintendent, estimating, scheduling, and cost engineering roles, on projects totaling $150 million in contract volume. She then served as national manager of construction administration for Blockbuster Entertainment's corporate design and construction team during the height of their development in the mid-to-late '90's. Fournier led departmental training, due diligence, construction contract approval, and audited cost controls for $100 million in annual design and construction spending, resulting in more than 350 new store openings per year across the United States and Canada.

In 1997 she was recruited by Miller Construction Company— a privately held design-build and general contracting firm located in South Florida—to lead revenue generation as vice president of business development, increasing market penetration in healthcare, higher education, telecommunications, and other mission-critical facilities. She purchased a minority stock position in the company and was later promoted to senior vice president and served on their board of directors through May 2004.

Expanding upon her experience in healthcare construction, she then went on to co-found a consulting firm to provide program management and owner's representation services for more than $50 million in hospital expansions, renovations, and other healthcare projects throughout the state of Florida.

Fournier's riveting stage presence and inspiring content draw thousands of attendees to a wide variety of industry events, webinars, and podcasts every year. Throughout her professional speaking career, she has performed with critical acclaim for chapters of the AGC, ABC, NAWIC, ASPE, and ACI, as well as The Blue Book Network Showcases, LinkedIn's Design & Construction Network, the Construction Industry Podcast, and many others. She is also a frequent guest lecturer on strategies for accelerated success to graduating construction management students at the University of Florida and other prestigious institutions nationwide.

Fournier's distinctive career has garnered her numerous industry awards, including: *The Business Journal's* Business Woman of the Year, *Fast Track* magazine's Top 40 under 40, *Heavy Equipment* magazine's Top 100 Construction Professionals to Watch, and *Business Leaders* magazine's Movers & Shakers Award. Fournier is a NAWIC scholarship recipient and a summa cum laude graduate of Michigan State University with a B.Sc. degree in building construction management.

Coty lives in Fort Lauderdale, Florida, with her daughter, Brielle Nicole. She loves the performing arts, music, books, movies and documentaries—and she's a passionate, lifelong student of theoretical physics, particle physics, cosmology and advanced mathematics.

CPSIA information can be obtained
at www.ICGtesting.com
Printed in the USA
FFOW01n1328151214

9555FF